Hybrid-Powered Brain

Takumi Satoh

Hybrid-Powered Brain

Neuron World Empowered by Ketone Bodies

 Springer

Takumi Satoh
Tokyo University of Technology
Hachioji, Tokyo, Japan

ISBN 978-3-031-54149-0   ISBN 978-3-031-54150-6   (eBook)
https://doi.org/10.1007/978-3-031-54150-6

This Springer imprint is published by the registered company Springer Nature Switzerland AG
The registered company address is: Gewerbestrasse 11, 6330 Cham, Switzerland

If disposing of this product, please recycle the paper.

This book is dedicated to my wife, Chiaki Satoh.

I have been happy just because She is with me.

Preface

Days of Kyoto University

I was a graduate student at Kyoto University [1988–1992] when I saw the then-President of the USA, George H.W. Bush on his way to sightseeing the Kyoto Imperial Palace in 1992. Then, I lived in an apartment in the Sakyo Ward and remember that the host lady always complained about her son's eating habits. He had just a cup of coffee for breakfast and had some meat for dinner. She worried continuously about his health and wished that her son ate carbohydrates, such as rice, in the morning.

Since she always took care of me, I wished to solve her worries. The next weekend, I went to her house after having bought a book written by a famous doctor. Opening the lattice door, I said to her "Breakfast without carbohydrate seems harmful to all persons. It is glucose that is supplied as an energy substrate to the brain. It is highly advisable to have rice or bread in the morning. This is a basic condition for the brain to have high performance. In addition, this is a comment of a famous person." She gently stroked the book with her hands, wearing a large green–blue opal ring, and said "Thank you so much. I will certainly read this book." This is one of the most impressive memories with my benefactors.

Not only the author of that book, but most people all over the world, believed that the brain does not work without glucose. This superstition has spread widely back then, and even now many people believe this. However, 30 years passed since this time. At that time, we had to change the code of the fixed telephone to send mail, but now an e-mail can be sent from the train within several seconds by hand unless you worry about hitting the face of the next person. Thus, we have come to a time when this common sense has to be revised from "glucose is the only source of brain energy." We must repaint the wall before big cracks, namely "common sense is only correct if blood glucose levels are stable."

Days of Neuroscientists

I will return to the story here when I was young. After I completed graduate school at Kyoto University and began to work in a pharmaceutical company, I realized that I was not a business person. Many friends had warned me that I was not a suitable businessperson, but 3 years had passed since I realized this. At last, I decided to return to being a researcher and searched for a position. I soon met Prof. Hiroshi Hatanaka at the Graduate School of Osaka University. He was then a famous professor of neuroscience in Japan and had completed writing the book "Brain as Material." When I met him, I took the first step toward becoming a neuroscientist. Since then, I have studied neuroscience for over 25 years. I have been walking with neurons in the brain. This is the system by which we can understand, by which we see and hear, by which we can remember what we think and feel, and by which we can choose what we like. The most interesting thing I ever felt was the system of finalizing life, named "neuronal death."

I moved to my present position 10 years ago, where I was determined to study the antiaging of the brain in terms of energy substrate as an emerging topic. At that time, I had the impression that few researchers were dealing with this topic. Ten years have passed, and I am writing this book. After 30 years as a researcher, my desk is full of information obtained from papers and books on neurons. In addition, I have a wealth of information from papers and books studying antiaging. They are mixed and fermented; some are simmering. Now, using gloves, let's pick out the bundle of paper processed to a heat-resistant temperature of 120 °C.

The following concepts emerged from the "steam."

"When blood glucose is not stable, neurons will be energy deficient. In the worst case, neurons will die. At least, they will not have their usual performance. The power to eliminate waste and impurities will decrease. The present problems of depression and Alzheimer's disease should be considered in terms of energy deficiency."

"If left unattended, the number of synapses, which are required for communication between neurons, will decrease. However, people can deal with energy deficiency on their own without dependence on any drug."

"When the brain cannot use glucose, it can function by using ketone bodies as an energy fuel. With this alternative, neurons can achieve much better performance, greater brain blood flow, greater capacity for clearance, and more efficient receptor systems."

Why I Wrote This Book?

I want to keep these boiled paper bundles properly before they disappear beyond oblivion. This is why I wish to publish this book worldwide. I will have less than 5 years before I retire from this position. A compound named "ketone bodies," which now are beginning to attract people's attention, are now facing the danger of being completely forgotten although ketone bodies, a group of compounds, are one of the most important energy fuels for the human brain.

I chose the two issues as the main topics of the book from the scientific point of view.

1. It is highly advisable to deal with the brain diseases of depression and dementia when the symptoms remain within energy deficiency. It is certainly possible to keep the progression much slower than the issues are left unattended. The one

energy substrate is certainly glucose, and the others are ketone bodies. If we properly use both glucose and ketone bodies, it is helpful to keep the brain at high performance for an extended period. For people to understand the brain energy problem properly, my possible task is to provide simple explanations of the brain energy system and to show the direction of eating habits for supplying brain fuel.

2. Big brain per body size is the most striking physiological feature of human beings. Why do they have so big brains? There are many possible answers to this issue according to their expertise. However, as far as I know, there is no available answer to what energy-producing systems supply the developing brain. Big names in biology must be thinking on this issue, but no person has reached a simple answer. By considering these backgrounds, I will try to provide you with a simple and scientific explanation of the following issues. By focusing on these two issues, I have determined to explain the important roles of glucose and ketone bodies.

(a) Why did human beings come to have a big brain?
(b) What energy system supply developing brain energy fuel? Which is more important for the brain, glucose or ketone bodies?

I am sure that most people, who will take this book, can read this book to the last page since I arranged the book so that readers can understand hypothesis as hypothesis and fact as fact.

Hachioji, Japan   Takumi Satoh

Goal of the Book

The world has been experiencing a great paradigm shift. We are shifting from "Expanding Society" to "Sustainable Living." However, we have not innovated yet a simple methodology to allow the brain to be sustainable and work well for our lifetime. "Sustainable Brain" is one of the most important elements of "Sustainable Living." Here I am going to propose a Practice for "Sustainable Brain." This book is aimed at "Sustainable Brain" by hybrid power of glucose and ketone bodies. Small Ketogenic is the most essential key to starting up this hybrid-powered brain.

Summary of the Book

To maintain positive feelings or sustain a fulfilling mood throughout the day, the priority is to send energy substrate to the brain. Since neurons have high energy demand, they are affected by energy shortage. The hybrid system of glucose and ketone bodies is the most effective method to preserve cognitive functions at high levels. In addition, hybrid power to the brain may help to maintain good cognitive function in older people. This is termed a "small ketogenic," which induces a slight increase in ketone bodies (0.2–0.5 mM). People can easily maintain the daily concentrations of ketone bodies by slight changes in eating habits. For example, it may be necessary to extend intervals between meals and allow the feeling of slight hunger to develop. This range of concentrations of ketone bodies is sufficient for keeping the brain calm and constitutively active.

Highlight 1: Hybrid-Powered Brain

To keep a positive feeling or spend the day in a fulfilling mood, the priority is to send out energy substrate to the brain. The active brain requires a supply of enough energy substrate to the brain. Neurons full of energy allow your feelings to become sharp, and your judgment is clear. On the contrary, how is the brain working under energy shortage? Since neurons have high energy demand, neurons are greatly influenced by energy shortage. Especially, low blood glucose may cause serious effects. When blood glucose levels are moving up and down in daily life, the serious problem is not high glucose, but low blood glucose because the brain is exposed to energy shortage. The brain finds it difficult to open the memory drawer. Feelings sink, and anger and sadness grow stronger. Everything may start to flow backward. To avoid these situations, it is worth considering the brain energy system that you don't usually care about and creating a chance of thinking about your brain.

Highlight 2: Small Ketogenic

Homo sapiens has two systems of brain energy. Glucose is produced from carbohydrates and ketone bodies are from fat. The two systems can compensate for each other. It is very easy to start up this hybrid system. You may have to extend the interval between meals and have time to feel small hunger in daily life within a reasonable range. This is called "Small Ketogenic." The hybrid system of glucose and ketone bodies is the most effective method to preserve cognitive functions at high levels. The hybrid-powered brain may contribute to the mentality of the old people in the village of longevity. They are enjoying daily life and farm working under these mentalities. The target range of concentrations of ketone bodies may be 0.2–0.5 mM. The range of concentrations of ketone body is sufficient for keeping the brain calm and constitutively active. We can attain healthy longevity by improving eating habits to very slightly (0.2–0.5 mM) termed as "Small Ketogenic." The big advantage of "Small Ketogenic" is to continue as long as we are alive. This eating habit does increase ketone bodies, but the increase is very small. The possible solution of ketone body concentrations (0.2–0.5 mM) is that anybody easily keep every day. This is why the author assumes that the Small Ketogenic (0.2–0.5 mM ketone body) empowers the brain and induces sufficient health effects.

Tutorial to Reader

It is an author's honor if you feel more familiar with your brain for high performance when you read this book. While reading this book, I am confident that surprising functions of the human brain may attract your brain.

1. All of the materials I have mentioned on the functions and structures of the brain are based on scientific papers.
2. There are many cases where "parables" and "personal impressions" are included in the text and enclosed within parentheses. These sentences shown by "( )" are inserted because scientific and formal statements can be easily understood. When you do not need to read these informal paragraphs enclosed within parentheses, these can be skipped.
3. Please feel free to read the *"Attention"* (a trivial with a cartoon) at the end of *Chapters 1–5.*
4. *References* are added to show scientific papers for readers to easily access to origins of knowledge. References are marked as "[ ]" such as [51] and [45] in the *Text*, *Figs*, and *Tables*. Please refer to these documents if you are interested in the original scientific data.
5. The main topic of this book is neuroscience but not eating habits. Thus, information on eating habits is highly limited. By using this book, I would like you to reconsider your brain energy for high performance in your work.
6. Ketone bodies, comprising three compounds (3-hydroxybutyrate (3HB), acetoacetate, and acetone), are considered alternative energy substrates when glucose is unavailable to the tissues. Since 3HB occupies over 85% of ketone bodies in the systemic circulation, 3HB may be described as "a ketone body." I will use both ketone bodies and 3HB in this book. In the physiological context, "ketone bodies" are often used. In addition, "3HB" is often used in the pharmacological context.
7. Since I am not a native speaker of English, plain English is used here.

Acknowledgment

Dr. Koujiro Tohyama, a former professor at Iwate Medical College, is appreciated with much respect and deep gratitude as he gave me many valuable photographs and clear neuroanatomical forward comments (*Figs. 1.2, 1.3, 1.4, 1.6, 1.7, 1.8, 1.9, 1.13, 5.1, 5.2, and 6.6*). It is my good luck to have his specialist input. He continues to research in the field of neuroanatomy at a high level. In particular, the first chapter would not be possible without his details.

My wife has always been with me and has provided me with many valuable comments. I desired to write clear comments in this book, but she warned me that the comments should be in a moderate tone. In addition, she asked detailed questions as a representative of the readers about the explanations that I thought were OK. Although I thought it was troublesome, many comments have now become understandable, so that readers can easily grasp the essence of their meaning.

Mr. Fumiyuki Minami presented illustrations of unique characters that added punch to the dishes finalized by cooperation. All characters are required, although the character that I like best is Miss Betz. I am grateful to the illustrator and like all of his characters.

Of course, I have unspeakable gratitude for Mr. Alexis Rivas (Springer), the editor of this book. This book can be published after a meeting with an excellent editor. He caught me at the tip of the antenna with that keen sensitivity and approved the publishing of a book.

Finally, I am very grateful to a certain person in the USA. This person is Prof. Stuart A. Lipton, who lives in La Jolla, California, a place famous for surfing and a wonderful town with many friendly people. He was running a large laboratory (Del E. Webb Center for Neuroscience) at the Burnham Institute, one of the top 10 institutes of all facilities in the USA. (Now he is a professor at the Institute for Translational Research in the Scripps Research Institute, right next door to the Burnham.)

Twenty years ago, while pursuing the identification of a receptor of NEPP11, a synthetic probe, I wrote an email to Stuart. Back then, he was studying the biological significance of S-nitrosylation, a mode of protein modification. He is known as one of the individuals who discovered memantine, one of the most famous antidementia drugs. At the time, I contacted a potential Nobel Prize candidate without thinking too much. Five minutes later, I got an answer, and 6 days later, I was alone standing at the Narita Airport to visit La Jolla for an interview. Then, I stayed in his laboratory for over 2 years. I was embraced by his considerably speedy English and had to adjust to his tight experimental schedule. He lived in a big mansion with his family and valued time with them. He returned home at a fixed time in the evening. Although I visited his office to discuss my ideas, he appeared not to listen to my uncertain guesswork and looked as if he was not interested in it. He never talked about wishful thinking or vague predictions. Just for that, I did not see the moment the light shone in his eyes and his mouth became relaxed as he talked. Several months later, we published a coauthored paper in *Proceedings of the National Academy of Science*. He was always working hard and deep in thought about something. But when I took my family to his lab, he was smiling and cheerful, like another person, and that is one of my modest memories of La Jolla. I learned my passion for research from Stuart. Thanks to him, I can talk about the surprising functions of the brain.

We were introduced to the wonderful daily life in La Jolla by him. My and my family's numerous pyramidal neurons of the neocortex have been deeply engrained with uncountable memories of relaxing time on the terrace of cafés on a Sunday morning and the vivid, colorful bundles of flowers that are always for sale in the morning market. To preserve these valuable memories until I am dead, I wish to continue the eating habit that activates the hybrid system of glucose and ketone bodies for the brain. My wish is that all the cells in my brain would preserve the memories of the seven-colored sky and sea that we saw at the Del Mar beach and the experience of California that we had, such as other small memories, including the smell of soy sauce.

Contents

Chapter 1
Neuron World

Abstract Amazing neuron world members!

Neurons are a group of brain cells. He is a superstar of all cells, which has a special ability. Various enzymes and mitochondria are moving and working secretly inside neurons. Each has a special ability and plays an important role. In addition, various types of glia are doing their jobs lively and faithfully to the missions to support neurons. Characters with unique features both inside and outside neurons are working together to form a wonderful world, whose capacity is far beyond computer systems. These neurons and glia are working crowded together inside your brain when you are reading this book now. This wonderful world can be called a "Neuron World" and each inhabitant will be introduced here in this chapter. Let's start slowly on the topics of glia (supporters) which are surrounding neurons.

1.1 Members of the Neuron World

1.1.1 Structure and Function of Neurons

Neurons are the most famous type of cells in the brain (Fig. 1.1). The structure and function of neurons are described in biology textbooks for high school students and may be familiar to most people. They have a critical function, acting as a switch for "information railroads," receiving information from other neurons as an input and extending information as an output. In addition, they participate in information storage and can act as a library, generating memories by fixing the information as a circuit. They have a very special structure, a soma, and a single, thick long protrusion for output, termed an "axon." Many oligodendrocytes, a complicated name that is difficult to pronounce, are stretched around the axon, enabling rapid exchange of information throughout the axon. In addition, neurons have a large number of short and thin protrusions called dendrites. This is an apparatus for input from other neurons. Neurons receive information from these sites. Neurons are connected to other neurons by special structures known as synapses; their actual structures can be seen under electron microscopy. They exchange information through various chemical compounds known as neurotransmitters, for example, acetylcholine and adrenalin. However, they are glia, not neurons, which occupy the majority of cells in the human brain. Other cell types are endothelial cells, smooth muscle cells, and immune cells.

Gilas have various roles that support the function of neurons. They can be classified into one of the following three groups (Fig. 1.1). This book will focus on brain function in terms of energy metabolism [1]:

1. Oligodendrocytes [2]
2. Astrocytes [3]
3. Microglia [4]

Oligodendrocytes act as complete electrical insulation, similar to plastic for electric wires, by wrapping around the axons of neurons multiple times. This wrapping increases the velocity of information transduction by several folds.

Astrocytes produce factors critically involved in the energy metabolism of neurons and supply energy substrates such as lactate and ketone bodies, which enable neurons to produce the "energy currency" of Adenosine TriPhosphate (ATP). For

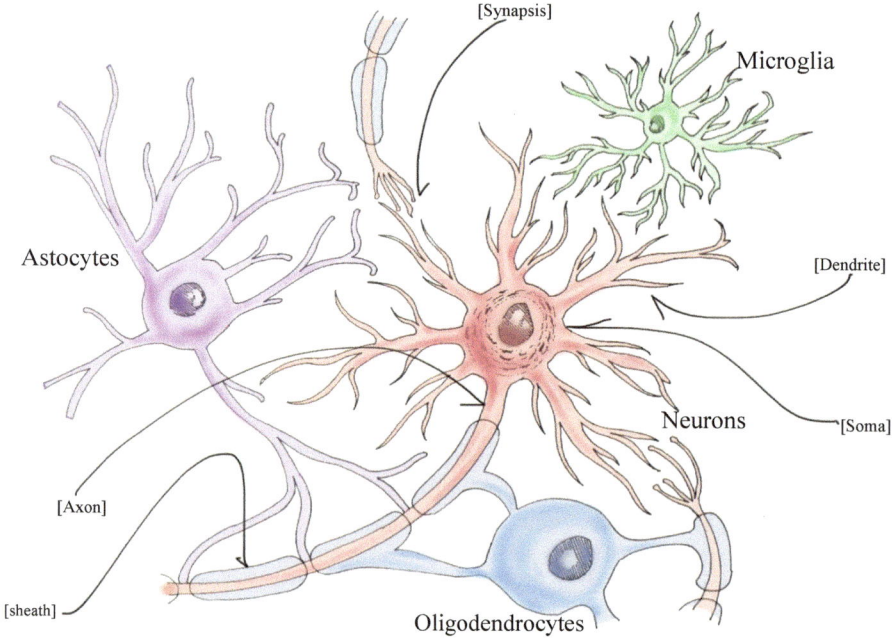

Fig. 1.1 Members of neuron world. Note that glia (astrocytes, microglia, and oligodendrocytes) are supporting neurons in every aspect. Neurons have highly specialized abilities dedicated to communicating with other neurons. However, the performances of neurons are possible by effective support by glia. This book will focus on the various interactions between neurons and astrocytes in terms of energy metabolism

neurons, the presence of astrocytes is like having a fuel station everywhere for a car. In addition, astrocytes can rapidly scavenge glutamate after the receptor activation. Astrocytes can be considered a "recycling center," where the scavenged glutamate is quickly reused for the next transmission.

It was believed that Microglia are not seen in healthy brains; they appear in response to inflammation and have the critical role of removing dead neurons by phagocytosis, similar to macrophages [5]. Their function is very important, similar to how a funeral home cares for the dead and in which microglia save neurons from emerge (Fig. 1.2).

However, the latest research suggests that different interpretations are possible. When neurons are partially injured, microglia can help neurons; when badly damaged, microglia can remove the neurons. Microglia not only process dead cells but also help neurons to recover. After all, they help to keep neurons healthy by acting as a "doctor" to injured neurons or removing dead neurons; they are "super-helpers" in the brain. It was first thought that microglia were not present in the healthy brain, but many microglia have now been identified in the healthy brain. As an aside, why did earlier researchers overlook the cells? It may be because they mistook them for another type of glia. Microglia are present throughout the whole brain in regular intervals, as shown in Fig. 1.3. Since many properties and functions of microglia remain unclear, I look forward to progress in this area in the future.

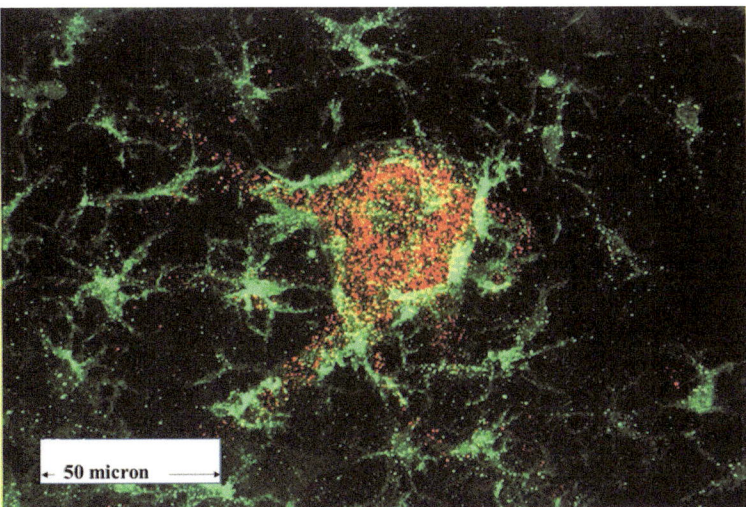

Fig. 1.2 Functions of microglia, Red: Damaged neuron, Green: Activated microglia. One micron corresponds to one thousand of a millimeter. Please note that neurons are large about 80 micron although microglia are small about 20 micron. Microglia may repair damaged neurons. By 20 years ago, it had been believed that microglia damaged neurons. Therefore, our view of microglia has changed 180 degrees

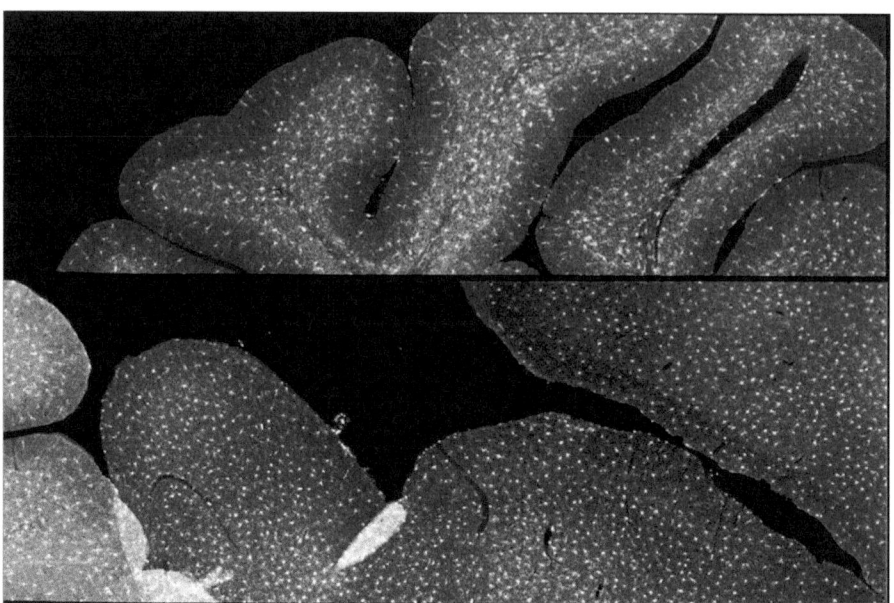

Fig. 1.3 Distribution of microglia. Microglia are present throughout the brain. Even now, microglia are the most mysterious cells in the brain. Microglia always have basic conflicts as to whether they are protective or destructive to neurons

1.1.2 Astrocytes and Blood-Brain Barrier

Now, we will take a more detailed look into the aforementioned astrocytes and oligodendrocytes. Figure 1.4a shows astrocytes (in green) in a section of the brain called the neocortex. This is the evolutionarily newest part of the cerebral cortex and occupies the largest portion of the human brain. From the perspective of anatomy, the term "neocortex" can be used here. Astrocytes are spread throughout the brain; in these images, they resemble the perfect starry sky. In addition, they connect by extending their protrusions like neurons. Figure 1.4b shows a microvessel, surrounded without gaps by nearby microcytes. Astrocytes receive nutrients and oxygen from microvessels and give the metabolites to neurons. (Previously, I stated that astrocytes are like gas stations, but they are more like a doorstep supplier, providing a more personalized service.)

The brain has a system, known as the blood-brain barrier (BBB), which I will mention here. This elaborate mechanism prevents most substances from entering the brain. Since only a highly limited number of compounds can enter the brain, you may get the impression that the BBB is impenetrable, but it is a very helpful system. Without the BBB, the brain would not be safe for a single day; that is how delicate, precise, and rigid the brain is. Passing through the BBB is the biggest challenge for drugs targeted to the brain. This is a major difficulty facing researchers developing drugs against dementia. A breakthrough in this area is a necessity; however, it is a

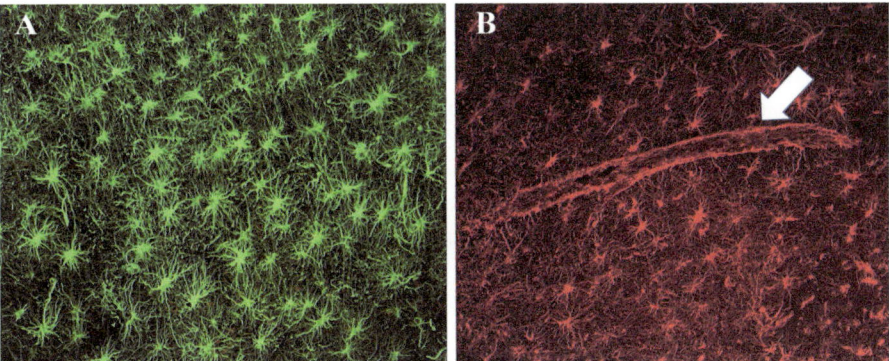

Fig. 1.4 Astrocytes are distributed to the brain and connected with other astrocytes (**a**). Astrocytes are communicating with each other as well as neurons do. Protrusions of astrocytes are wrapping all microvessels in the brain (**b**). This wrapping may provide roles of the blood-brain barrier. No molecules including glucose and 3HB cannot target neurons without passing through astrocytes. Astrocytes occupy all of the gaps between neurons. Therefore, all compounds target neurons by the following sequences: microvessels-astrocytes-neurons. There must be no exceptions

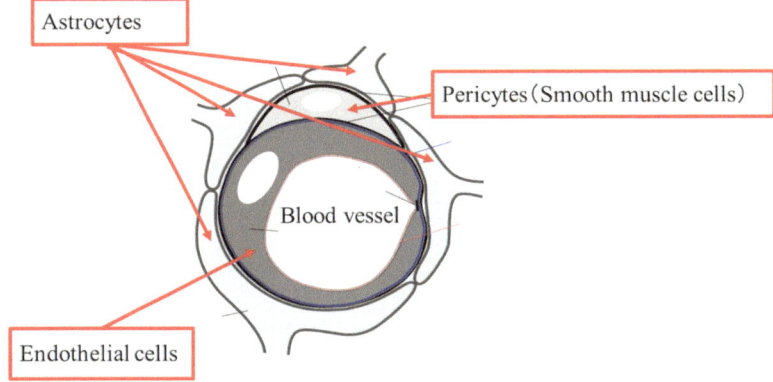

Fig. 1.5 Structure of blood-brain barrier. All microvessels of the brain are surrounded by protrusions of astrocytes [6]. Thus, this system strictly protects the human brain against exogenous compounds. Since protrusion of astrocytes is wrapping all microvessels of the brain, astrocytes may determine which compounds can pass through the blood-brain barrier. Glucose and ketone bodies are selected as special compounds that have a free path by astrocytes. Glucose and ketone bodies have a big advantage over other chemical compounds that affect brain functions

complicated problem. On the other hand, the biggest advantage of ketone bodies, which I will talk about in this book, is the high ability to pass through the BBB. In contrast, most synthetic drugs cannot pass through and a very small number of them may have very low ability.

In the BBB, endothelial cells, pericytes (a type of vascular smooth muscle cells), and astrocytes work together to enable selective permeability, as shown in Fig. 1.5. BBB can specifically select compounds that enter the brain from the various components of blood.

The reason why BBB can be seen only in the brain is that microvessels are wrapped several times by protrusions extending from astrocytes as shown in Fig. 1.5. The functions of endothelial cells are not different from those of other tissues. Since brain microvessels are wrapped by the protrusions of astrocytes, selective permeability may be possible (Fig. 1.5). Protrusions of astrocytes do not explain the selective permeability of BBB, but they are likely to play an essential role.

In particular, when the brain selectively takes up glucose and ketone bodies as energy substrates, the roles of the protrusions are highly essential. They can function as relay stations for the delivery of glucose and ketone bodies (and in some cases their metabolites) from microvessels to neurons. Energy substrates in the microvessels can be delivered to neurons through the protrusions of astrocytes after passing through the vessel endothelium. Consequently, has the true identity, real face, and true face of the blood-brain barrier come into view a little? However, the true face of the BBB remains to be examined [7]. Although I have some other things to describe, different perspectives are important, so I will end this topic here [8].

1.1.3 Astrocytes and Synaptic Transmission

I will continue on the topic of glia. Astrocytes are necessary for synaptic transmission (Fig. 1.6). Notably, astrocytes surround synapses with no gaps. Most of the synapses are the same and this is the reason that astrocytes are necessary for synaptic transmission. It is well known that neurotransmitters are released into the synaptic gap from synaptic vesicles removed by astrocytes for reuse. Various problems

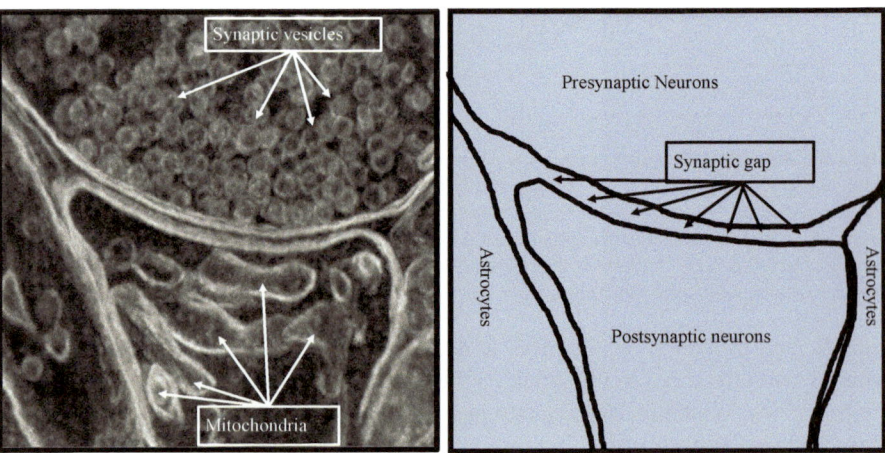

Fig. 1.6 Essential roles of astrocytes for synapse formation. Note that all synapses are surrounded by astrocytes. Astrocytes are involved in neurotransmission as well as energy supply. It had been believed that astrocytes uptake released neurotransmitters. However, recent studies suggest that astrocytes are directly involved in synaptic transmission between neurons

will occur if they stay there unchanged. Synaptic transmission cannot proceed further without astrocytes [9].

By considering these facts, the BBB incorporates nutrients and energy as well as synapses that convey information as relay points do not work anymore without the assistance of astrocytes. Astrocytes play a supporting role in the complex systems of the brain and may be a critical part of good brain function [6–8].

1.1.4   Oligodendrocytes as an Insulator of Axons

Finally, we discuss oligodendrocytes (Fig. 1.7). As shown in the left image, there are many unstained bundles of lateral lines. These must be bundles of neuronal axons. The black-stained protrusions are oligodendrocytes, which warp neuronal axons several times without a gap. All of the protrusions of oligodendrocytes wrapping neuronal axons are so thin that immunological signals are just seen obscure [10].

More than half of the wrapping membrane is cholesterol. Thus, cholesterol is an important chemical in the brain. As I mentioned earlier, neuronal information can be transduced rapidly because oligodendrocytes stretch their protrusions to the plunge and wrap axons (Fig. 1.7b). They function in the same way as insulators of electricity and allow neurons to send out electric signals. Briefly, I described three types of glia. Each cell works hard and steadily to complete its function. (Glia are "super supporters" and "global escorts" that support neurons; isn't this encouraging? Next, we will discuss neurons, a cell type with a prince- or princess-like existence.)

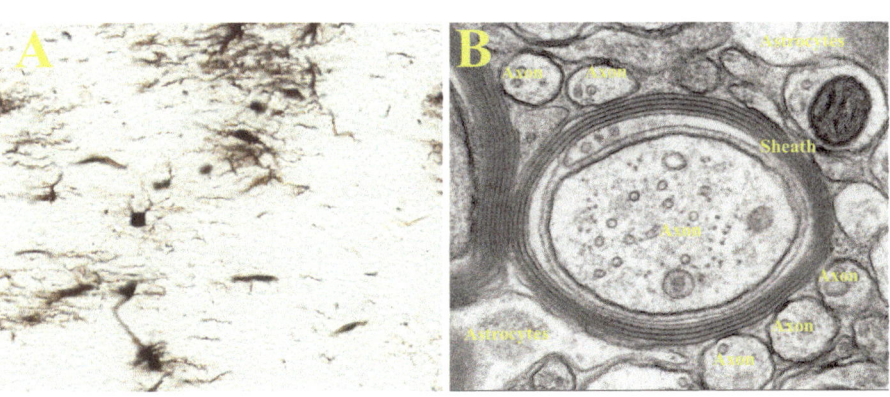

There are axons in the lateral directions (Optical microscopy).   Oligodendrocytes wrapping around axons (Electron microscopy).

Fig. 1.7 Distribution of oligodendrocytes (**a**). Why are all of the oligodendrocytes not stained black even though they are wrapped several times by axon? This is because the protrusions of oligodendrocytes repeatedly wrap axons so thinly that the contents of the cytosol are almost not stained. Function of oligodendrocytes (**b**). Oligodendrocytes are wrapping neuronal axons for insulation. The content of cholesterol is rather high in the cell membrane of oligodendrocytes. This is why cholesterol has a big impact on brain functions

1.2   Two Types of Neurons

1.2.1   Basic Structures of Neurons

It is said that the neocortex contains 20 billion neurons. (Information on the number of cells is constantly being revised.) Figure 1.8 shows the typical morphology of neurons. A neuron is composed of three parts [12].

1. Dendrites, to receive input from other neurons
2. Soma, including the nucleus
3. Axons, for output to other neurons

In brief, the functionality of neurons can be divided into three parts; the dendrites receive information, the soma integrates information, and axons send out information (Fig. 1.8a). In addition, numerous small on-grain structures of the dendrite are called the spine (Fig. 1.8b). Each spine is considered a synapse. The synapse is a

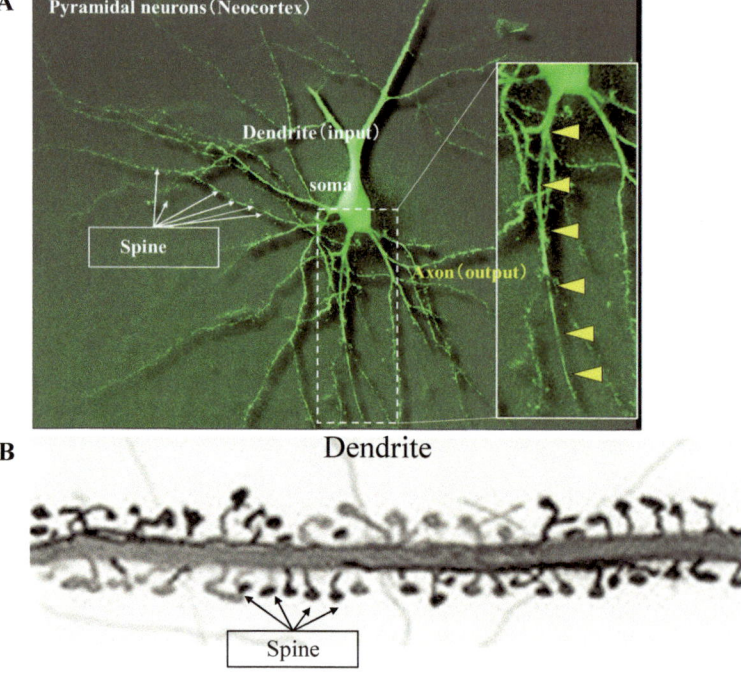

Fig. 1.8 Basic structure of a neuron (**a**). A single neuron has many dendrites for informational input and a single axon for informational output. A spine on dendrites is corresponding to a single synapse. This book will focus on pyramidal neurons because of their vulnerability to energy shortages such as fluctuation of blood glucose levels. Schematic structure of spines on a dendrite (**b**). A single dendrite has thousands of spines, which correspond to postsynaptic structures of excitatory neurons to receive neuronal information from other neurons [11]. Structural degenerations of pyramidal neurons are initiated by the decline of spines on dendrites during the progression of dementia

small apparatus for the transmission of information, similar to semiconductors in a computer. Each neuron has thousands of spines. The number of pieces of information that pass through the brain is astronomical. For example, during a walk in the forest or the park, as many signals are exchanged in the brain as there are stars in the clear night sky; alternatively, while you are chatting and having a cup of tea, a milky way of neuronal signals may be formed in the brain.

1.2.2 *Large Pyramidal Neurons and Small Granular Neurons*

There are two types of neurons. The larger ones are pyramidal neurons, and the smaller ones are granular neurons. Figure 1.9 shows pyramidal neurons in the neocortex. The white arrowed cells have a large pyramid shape.

In general, the cell sizes are as below:

Betz pyramidal neurons are over 100 microns.
Pyramidal neurons are approximately 20–50 microns.

One micron corresponds to one thousand of a millimeter. (It seems that 10 Betz pyramidal neurons and 20–50 pyramidal neurons would line up to a 1 mm scale of the ruler. In addition, 50–100 granule neurons line up.) Betz giant pyramidal neurons are motor neurons that extend axons from the neocortex to the skeletal muscle of hands and legs (or nearby). The longest axons are extended to the waist. For

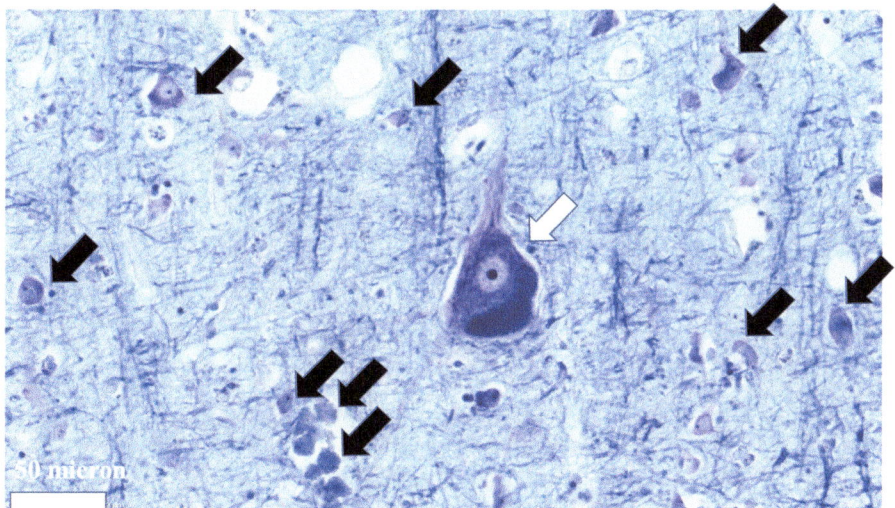

Fig. 1.9 Neurons in layer IV of the neocortex, Betz giant pyramidal neurons (white arrow) and pyramidal neurons (black arrow). Note that the Betz pyramidal neuron is extraordinarily big compared to other pyramidal neurons. Betz giant pyramidal neurons are special for their large volume from other pyramidal neurons

emphasis, I will state again; that the axons extend from the neocortex to the waist. If Betz's giant pyramidal neurons were expanded to the same size as a human, the axons would be over 400 km. The size may be beyond our imagination, but these cells are huge and very long. It is really surprising! Of course, these are the biggest cells in humans if axons are included. What a straightforward and disciplined cell to reach out and give instructions on the spot that is effective by just a command. What a wonderful character the cell is!

Pyramidal neurons are a group of cells not only in the neocortex, but also in the hippocampus, cerebellum, and midbrain; they are everywhere in the brain. Pyramidal neurons serve as an information hub that connects with each network of the brain [12].

1.3 Pyramidal Neurons

1.3.1 Dementia and Decrease in Hippocampal Pyramidal Neurons

The neocortex has 20 billion neurons, but most of them are granular neurons and the others are pyramidal neurons. Figure 1.10 shows the division of labor between pyramidal and granular neurons. Pyramidal neurons accept a large amount of information from nearby granular neurons and sometimes keep them and sometimes send them out to other regions of the brain. Pyramidal neurons have a ten times larger

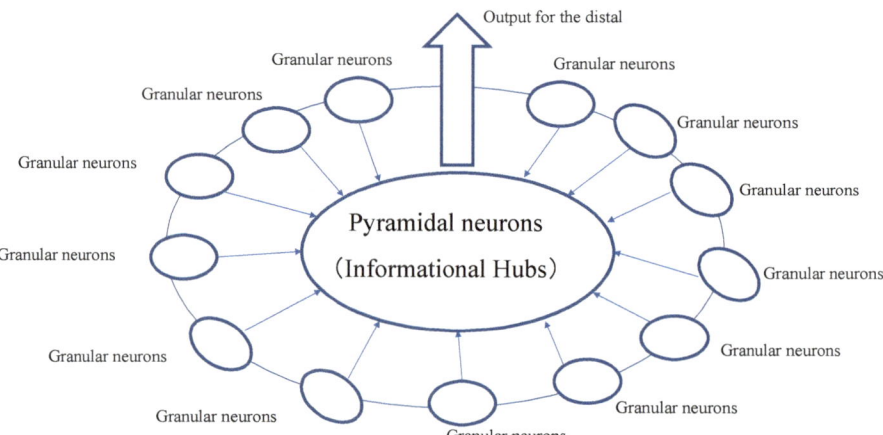

Fig. 1.10 Pyramidal neurons function as informational hubs that accept a huge amount of information from other granule neurons. Therefore, a breakdown of pyramidal neurons may cause serious damage to the whole brain. Since pyramidal neurons demand a larger amount of energy than granular cells, pyramidal neurons are subject to energy shortages. This causes serious effects during the early stage of dementia in the hippocampus

volume of granular neurons. Pyramidal neurons have many more roles than granular neurons. If pyramidal neurons are down, the brain may accept serious problems. (Please imagine the situation in which Haneda or Narita airport is shut down.)

Taking "dementia" (Chap. 6) and "epilepsy" (Chap. 11) as typical examples of brain disease, most of the debate concentrates on energy problems in pyramidal neurons. Because when the brain is not going well, it is the pyramidal neurons, not the granule neurons, that are the first to show the disorder.

The hippocampus will be in a more serious situation than the neocortex (Fig. 1.11). This is highly characterized by a densely crowded arrangement of each line of neurons. This is why there is a danger of losing many pyramidal neurons at the same time under potent stress. Since the hippocampus has a much smaller population than the neocortex, functionality will be increased by the neurons being crowded together. However, the brain is in danger of suddenly losing all of its hippocampal pyramidal neurons; this is both an advantage and a disadvantage [12]. This volatility of the hippocampal pyramidal neurons to stress is closely related to the progression of Alzheimer's disease (AD) and will be mentioned in Chap. 6.

As shown in the Magnetic Resonance Imaging (MRI) images in Fig. 1.12, many pyramidal neurons become lost during the progression of AD. The hippocampus is shown by a white oval (left: healthy person; right: patient with AD). The hippocampus is progressively degenerated in the brain of patients with AD, and the patient's symptoms will progress considerably. Since degeneration is also enhanced in the neocortex, the patient cannot retain short-term memory, but may also begin to lose the old, important memories. It is the hippocampus that first degenerates.

When neurons are supplied sufficient fuel, neurons have an input and output of information that works well, and information is smoothly exchanged. Your brain

Fig. 1.11 Hippocampus in the human brain. This region of the brain is highly specialized for saving short-term memories and is highly vulnerable to energy shortages. This book will concentrate on the energy problem of hippocampal pyramidal neurons during the early stages of dementia. Loss of ability to maintain short-term memories is a functional criterion for diagnosis of dementia

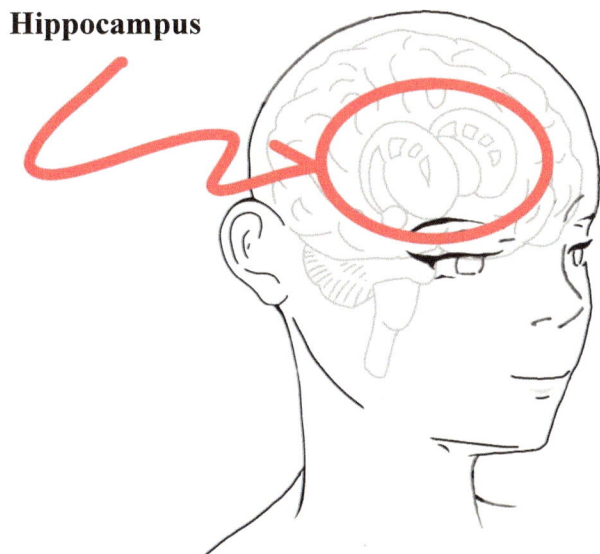

Hippocampus

A   **B**

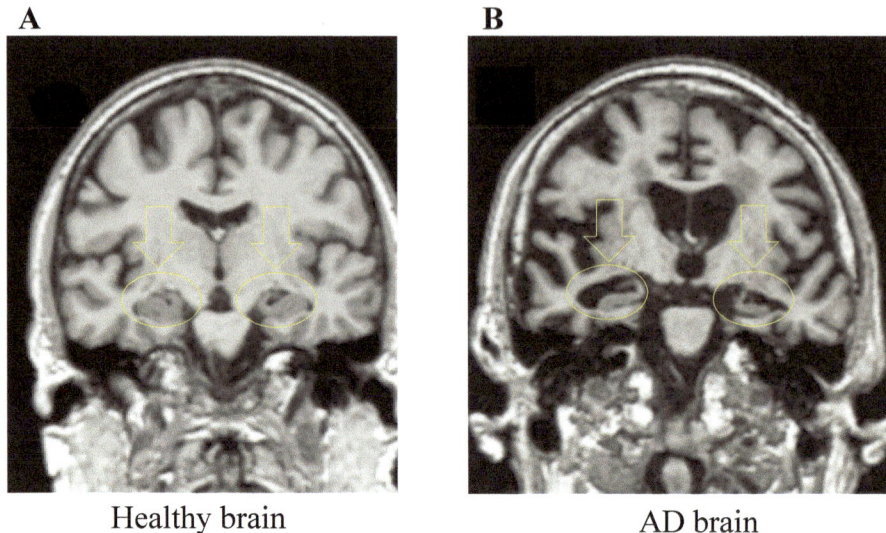

Healthy brain   AD brain

Fig. 1.12 MRI image of AD brain [13]. Circles show the hippocampus. Not that the AD brain has lost many neurons. The hippocampus of the AD brain has large spaces compared to the healthy brain. These spaces are due to the degeneration of neurons in the hippocampus. This is one of the most important criteria for diagnosis of AD

works well when the pyramidal neurons, of which there are numerous populations, are performing well. Previous experiences, acquired intelligence, remembered images, familiar melodies, memories of taste and smell, and all information are concentrated in your brain, so it may be possible to reach the best answer for you.

1.3.2 Importance of Energy Substrate Supply to the Brain

The high performance of pyramidal neurons is required for a healthy brain. You will live a hopeful life if your brain keeps active by concentrating on the maintenance of brain energy. From a neuroscientific perspective, brain activity is determined at least in part by serotonin, the "happiness" hormone in the brain. To realize an active brain, it is necessary to supply sufficient energy substrate to pyramidal neurons. (A potential subject for this book may be considered "Hope from brain energy systems.")

To maintain positive feelings or sustain a fulfilling mood throughout the day, the priority is to send energy substrate to pyramidal neurons. If pyramidal neurons are full of energy, your feelings become sharp, and your judgment is clear. Naturally, this does not mean that all of the problems of human characters and ways of thinking will not be solved. This is just one possible answer. However, it is worth trying [14].

1.3.3 Low Blood Glucose Is a Serious Problem for the Brain

In contrast, how does the brain work with an energy deficiency? Since pyramidal neurons have a high energy demand, pyramidal neurons are potently influenced by energy deficiency. In particular, low blood glucose may cause serious effects.

It is often said that glucose is "brain energy." This is the correct opinion. However, there is a single condition that blood glucose should be stable; however, stable blood glucose is not an easy task. In most people, blood glucose levels move up and down in daily life. Most people are worrying about high glucose, but not about low blood glucose, which leads to a serious problem. This is because pyramidal neurons are suffering from energy deficiency. It is a hassle to think, and the brain wants to get away from this hassle. It becomes difficult to access memories. Feelings sink, and anger and sadness grow stronger. Everything may start to feel backward.

To avoid these situations, the brain energy system, which is often ignored, should be considered, including pyramidal neurons. Books sometimes have the power to relate to the lives of readers by changing their feelings and ways of thinking. I would like to insert this power into a page of this book. My secret wish is to create a difference in the readers' daily lives.

1.3.4 The Brain Needs Ketone Bodies as Energy Substrates

Therefore, I propose again that glucose alone is NOT sufficient for the high performance of pyramidal neurons. I would like to make a statement that other energy substrates, ketone bodies, are required to keep pyramidal neurons working well. Of course, this is not new information. Some already know this. But many people, but not everyone, do not know this. Thus, I repeat this here.

Let us return to the topic of neurons. Figure 1.13 shows images of pyramidal neurons and granular neurons of the hippocampus. It is easy to see that pyramidal neurons have a volume at least ten times larger than that of granular neurons. When pyramidal neurons are, for example, 3 times larger in diameter, they have 27 times larger volume than granular neurons. This book will explore the energy supply system to neurons and would like to propose a method to preserve irreplaceable pyramidal neurons. Please read the next paragraph for a more interesting perspective.

Pyramidal neurons (CA1)   Granular neurons (dentate gyrus)

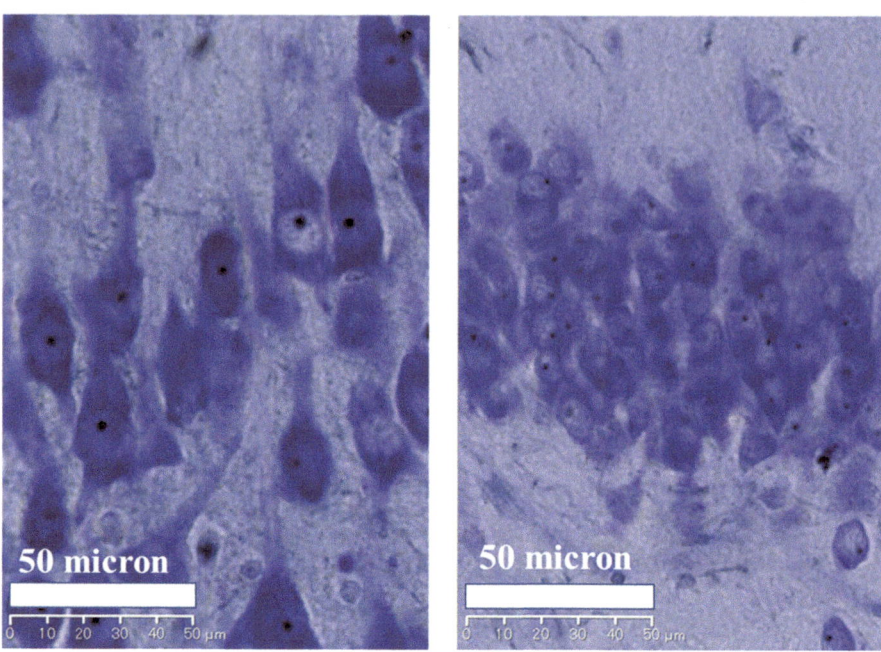

Fig. 1.13 Pyramidal and granular neurons in the hippocampus. Pyramidal neurons are much bigger than granular neurons and more sensitive to energy shortages

1.4   Roles of Astrocytes (Lactate Shuttle)

1.4.1   Astrocytes as Admirable, Earnest, and Hard Workers

Astrocytes are directly responsible for supplying energy to neurons, such as pyramidal neurons. Astrocytes must carry food without fail to the energy-hungry pyramidal neurons. This big eater has a great taste for food. In addition, non-neuronal cells can cook themselves when glucose is delivered. However, neurons do not feed themselves. They need special care.

Their favorite food is NOT glucose, which cannot be easily digested, but a substance that is digestible and provides a good energy source for neurons. This is lactate [3, 13]. An astrocyte is like a maid who delivers a pot containing lactate for neurons to enjoy and feel comfortable. In addition, an astrocyte allows neurons to drink lactate by cooking glucose, just as a babysitter gives a baby food. If the "maids" (astrocytes) do not deliver lactate to neurons, there is no doubt that neurons will become a complete "*treasure rot.*" Astrocyte is sensible, serious, and hardworking. Therefore, we may have to take care of astrocytes, not only of neurons. Although the three types of glia have distinctive and important roles, astrocytes are special in terms of the supply of energy substrate to neurons named as "*lactate shuttle.*" This

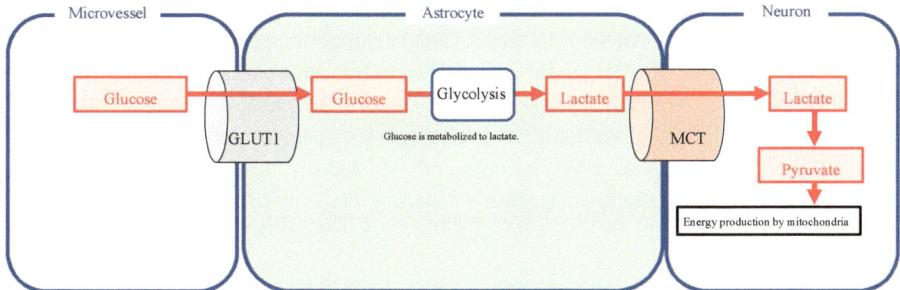

Fig. 1.14 Lactate shuttle accounts for 70% of the Energy supply for neurons [15, 16]. Note that astrocytes make lactate from glucose and donate to neurons through monocarboxylic acid transporter (MCT). This is the main energy of neurons. Therefore, neurons cannot utilize their specialized abilities without this lactate supply. Astrocytes are critical linkers between microvessels and neurons, suggesting the importance of the lactate shuttle by astrocytes

system is installed only in the brain. Astrocytes are responsible for the shuttle (Fig. 1.14).

Astrocytes play a central role in this lactate shuttle, serving as relay points between microvessels and neurons. For a simple analogy, imagine that astrocytes take the raw wood out of the box and quickly load them onto the conveyor belt, cut them into small pieces of charcoal, and throw them onto a nearby stove (neuronal mitochondria). Inside astrocytes, the following work is done. Glucose is phosphorylated twice, divided into two parts of C3 compounds, and converted to pyruvate by dehydrogenation. In addition, pyruvate is reduced to lactate. This process is completed within 1 s, but astrocytes perform this work continuously every day. They may be a tiny wizard [1].

1.4.2   Amazing Mechanism of Lactate Shuttle

So let us consider the lactate shuttle step by step. Glucose accounts for the majority of the energy substrate in the blood. Glucose enters astrocytes through GLUT1, a specific path for glucose on the cell membrane. By repeatedly wrapping on a microvessel, astrocytes can more effectively take up nutrients, including glucose, into the cytoplasm. This is part of the lactate shuttle. GLUT1 is a path for glucose without limitation. Usually, insulin can allow glucose to enter, but GLUT1 does not need the presence of insulin. When glucose levels are stable, astrocytes can steadily take up glucose. This is a highly excellent glucose uptake system when there is a stable supply of glucose. Glucose is metabolized into two molecules of lactate by glycolysis [15, 16].

Lactate produced in astrocytes can enter neurons through a specific path, termed the "monocarboxylic acid transporter (MCT)." Indeed, astrocytes and neurons each have an MCT since the two molecules are close enough to take a single path.

GLUT1 between astrocytes and neurons has the same story. Lactate and glucose never diffuse into the cerebrospinal fluid. Glucose can move to astrocytes, bit by bit, and lactate can move to neurons bit by bit. Please imagine a relay team that steadily passes the baton to the next runner with an underhand pass. The baton is handed over so quickly that we cannot catch the motion; for sure, they will win the race. In neurons, lactate is converted to pyruvate and oxidized in mitochondria to produce ATP, an intracellular energy substrate. Lactate can effectively accomplish its purpose.

1.4.3   Two Pathways for Glucose (GLUT1 and GLUT3)

Lactate, supplied to neurons through the lactate shuttle, is the most important energy substrate in neurons. The lactate shuttle accounts for 70% of the total energy consumption of neurons. The remaining 30% is supplied as glucose itself through GLUT3 (Fig. 1.15). Although this system occupies just 30% of the total energy supply, this is closely connected with a special function of evoking action potentials.

Attention 1.1!

Do you know that Mr. Oligodendrocyte upregulates motor skills? When a child develops motor skills by recombination of the neuronal circuits, an adult may find this difficult. However, even an adult can become a more skillful soccer player or learn a refined technique of table tennis by hard training. This is because Mr. Oligodendrocytes can remodel neuronal circuits. Although Mr. Oligodendrocyte increases the thickness and length of the sheath that wraps the neuronal axon, an action potential can be more rapidly communicated between neurons (Fig. 1.16). In

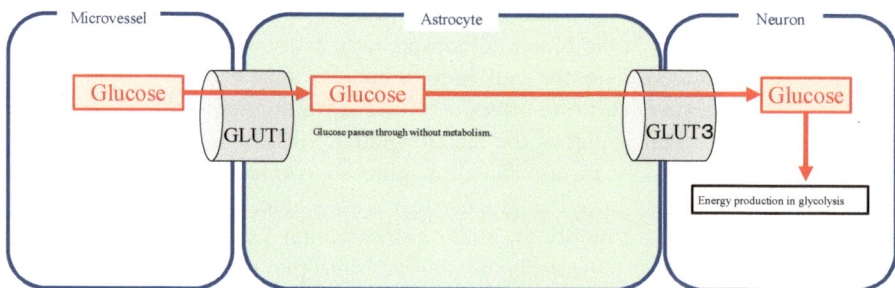

Fig. 1.15 Glucose uptake accounts for 30% of the energy supply for neurons [16]. The glucose is used for glycolysis for energy supply to the Na$^+$ pump. The energy of neurons is supplied as lactate (through the lactate shuttle) and 30% is supplied as glucose itself. Neurons have a big appetite for energy, so they need both (lactate and glucose)

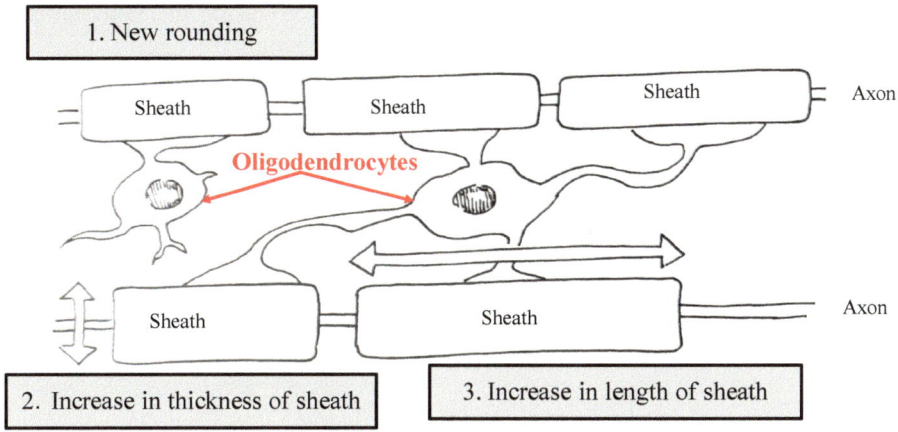

Fig. 1.16 Essential roles of oligodendrocytes for acquisition of motor skill [10]

addition, Mr. Oligodendrocyte himself produces cholesterol as a building block from acetyl-CoA. They use a lot of fat and sugar that I like, too, although I am a little concerned about this because it reduces my share of them. Well, who am I? I have been in cells for 2 billion years. I produce ATP, an energy substrate. I am a reliable fairy!

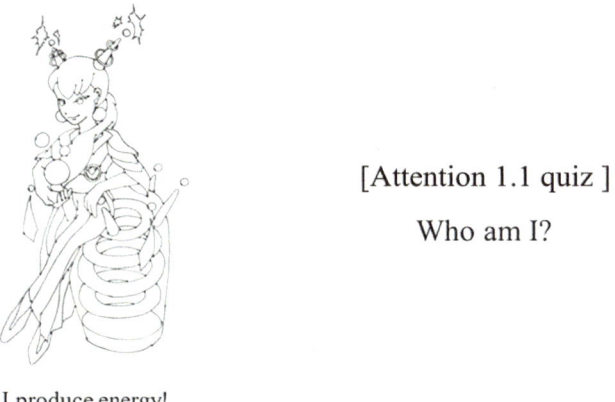

[Attention 1.1 quiz]

Who am I?

I produce energy!

Attention 1.2!

Today, I will talk about Betz giant pyramidal neurons, a big name among the big names of my relatives. If my Aunt Betz is taken below the Tokyo Station and expanded to human size, her axons could easily cross over the summit of Mt Fuji and reach Osaka Castle—they are that long! She is a superstar among our relatives.

We are called "neurons." We are small and numerous. By using the neuronal network, we send more and more information. Isn't it amazing?

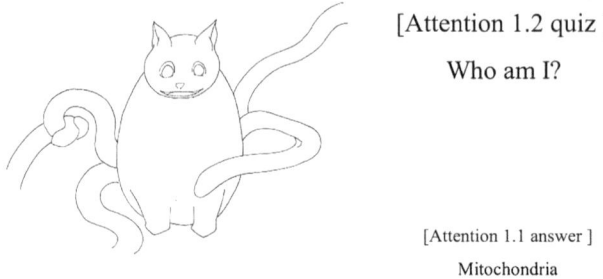

[Attention 1.2 quiz]

Who am I?

[Attention 1.1 answer]

Mitochondria

I am a small neuron in the brain.

Attention 1.3!

Nice to meet you for the first time.

Today's topic is the influence of the rise and fall of blood glucose. When glucose increases after carbohydrate ingestion, insulin rapidly increases. Then, blood glucose decreases rapidly in response to an insulin spike. Please take care that blood glucose does not decrease too much. Please watch your blood glucose levels by having coffee or tea without sugar when you enjoy a sweet cake. I am a person who takes care of neurons when they are damaged. I always take care of neurons. I am with neurons as well as astrocytes.

[Attention 1.3 quiz]

Who am I?

[Attention 1.2 answer]

Granular neuron

I rescue neurons in crisis.

Attention 1.4!

Today's topic is the MCT, the monocarboxylic acid transporter. Essentially, the MCT provides a path through the membrane just for carboxylic acid. When glucose is divided into two lactate molecules, lactate can pass through the MCT. Later, someone will give a detailed explanation of carboxylic acid. I am too busy to do it today. Goodbye! I will see you again. I am the largest cell in humans, extending from the brain to the legs and involved in motor function.

[Attention 1.4 quiz]

Who am I?

[Attention 1.3 answer]

Microglia

I am the biggest cell in the human body.

References

1. Magistretti PJ, Allaman I. A cellular perspective on brain energy metabolism and functional imaging. Neuron. 2015;86(4):883–901.
2. Elbaz B, Popko B. Molecular control of oligodendrocyte development. Trends Neurosci. 2019;42(4):263–77.
3. Ferguson BS, Rogatzki MJ, Goodwin ML, Kane DA, Rightmire Z, Gladden LB. Lactate metabolism: historical context, prior misinterpretations, and current understanding. Eur J Appl Physiol. 2018;118(4):691–728.
4. Guedes JR, Ferreira PA, Costa JM, Cardoso AL, Peça J. Microglia-dependent remodeling of neuronal circuits. J Neurochem. 2022;163(2):74–93.
5. Wolf SA, Boddeke HW, Kettenmann H. Microglia in physiology and disease. Annu Rev Physiol. 2017;79:619–43.
6. Abbott NJ, Rönnbäck L, Hansson E. Astrocyte-endothelial interactions at the blood-brain barrier. Nat Rev Neurosci. 2006;7(1):41–53.
7. Alexander JJ. The blood-brain barrier (BBB) and the complement landscape. Mol Immunol. 2018;102:26–31.
8. Segarra M, Aburto MR, Acker-Palmer A. Blood-brain barrier dynamics to maintain brain homeostasis. Trends Neurosci. 2021;44(5):393–405.
9. Blanco-Suárez E, Caldwell AL, Allen NJ. Role of astrocyte-synapse interactions in CNS disorders. J Physiol. 2017;595(6):1903–16.

10. McKenzie IA, Ohayon D, Li H, de Faria JP, Emery B, Tohyama K, Richardson WD. Motor skill learning requires active central myelination. Science. 2014;346(6207):318–22.
11. García-López P, García-Marín V, Freire M. Dendritic spines and development: towards a unifying model of spinogenesis—a present-day review of Cajal's histological slides and drawings. Neural Plast. 2010;2010:769207.
12. English DF, McKenzie S, Evans T, Kim K, Yoon E, Buzsáki G. Pyramidal cell-interneuron circuit architecture and dynamics in hippocampal networks. Neuron. 2017;96(2):505–520.e7.
13. Vemuri P, Jack CR Jr. Role of structural MRI in Alzheimer's disease. Alzheimers Res Ther. 2010;2(4):23.
14. Cunnane SC, Trushina E, Morland C, Prigione A, Casadesus G, Andrews ZB, Beal MF, Bergersen LH, Brinton RD, de la Monte S, Eckert A, Harvey J, Jeggo R, Jhamandas JH, Kann O, la Cour CM, Martin WF, Mithieux G, Moreira PI, Murphy MP, Nave KA, Nuriel T, Oliet SHR, Saudou F, Mattson MP, Swerdlow RH, Millan MJ. Brain energy rescue: an emerging therapeutic concept for neurodegenerative disorders of aging. Nat Rev Drug Discov. 2020;19(9):609–33.
15. Suzuki A, Stern SA, Bozdagi O, Huntley GW, Walker RH, Magistretti PJ, Alberini CM. Astrocyte-neuron lactate transport is required for long-term memory formation. Cell. 2011;144(5):810–23.
16. Brekke E, Morken TS, Sonnewald U. Glucose metabolism and astrocyte-neuron interactions in the neonatal brain. Neurochem Int. 2015;82:33–41.

Chapter 2
Brain Energy Problem

Abstract Until recently, it was common sense that the brain's energy is obtained only from glucose. However, accumulating evidence shows that glucose alone does not allow the brain to work well. The previous common sense is based on the following three facts:

1. The lactate shuttle accounts for 70% of the total energy supply of neurons [Bélanger et al. (Cell Metab. 14:724–738, 2011)].
2. Glucose uptake accounts for 30% [Dienel (Physiol Rev. 99:949–1045, 2019)].
3. Both are dependent on stable glucose levels [Xu et al. (Nature 556:505–509, 2018)].

Item number 3 leads to a serious problem. When the blood glucose level is not stable, that is, the glucose level increases and decreases, and neurons cannot prevent energy deficiency. Depression and dementia are caused by this. This is the time when an alternative energy substrate plays an important role. Now, the ketone bodies finally act here. Ketone bodies are actors in the leading role in this book. Glucose and ketone bodies are double main casts in this book. When the relay runner mentioned earlier cannot run due to injury, the reserve player will run in the match. Unexpectedly, he runs to the top. The ketone bodies play the roles of a super-substitution.

2.1 Brain Energy Substrates

Ketone bodies may not be required if blood glucose is stable. However, when blood glucose is unstable (actually, most people may have unstable blood glucose in their daily lives), the brain cannot work well without the assistance of ketone bodies. Chapter 5 explains how glucose and ketone bodies can reach neurons in the brain.

In a healthy person (when blood glucose levels are stable), energy demand may be satisfied with glucose, as many people believe. This hypothesis may be correct just in healthy people. Generally, one hypothesis cannot be applied to anyone, as every hypothesis has an exception. The brain of the person with fluctuating blood glucose may be suffering from energy deficiency without the assistance of ketone

T. Satoh, *Hybrid-Powered Brain*, https://doi.org/10.1007/978-3-031-54150-6_2

bodies. If you look into the symptoms of a patient with genetic mutations of the gene of glucose transporter such as GLUT1, you can easily understand that the "invisible" help from ketone bodies, such as 3-hydroxybutyrate (3HB), is essential for the brain. When they (GLUT1 deficiency) feel hungry, they are at risk of coma [1].

Indeed, if a person has the correct gene coding for the enzyme to synthesize ketone bodies, they may live a daily life free from the risk of coma. However, when too much carbohydrate is ingested, the enzyme for ketone body synthesis will be shut off as a result. Not least, this creates a highly dangerous situation for the brain. I do not recommend strict restriction of carbohydrates, but there is a need to avoid eating too much carbohydrate. This is critical to keep the brain healthy for a long time [2].

2.2 3-Hydroxybutyrate as an Energy for the Brain

If you are familiar with biochemistry, you can omit Sects. 2.2–2.4. However, those who have an interest in chemistry are encouraged to carefully read the sections. Please look at the chemical structure of the major compound of ketone bodies (3-hydroxy butyrate 3HB). Hydrogen (H) atoms bind to carbon (C) or oxygen (O) to form a single group (–OH) or (–OH). As mitochondria produce the intracellular energy currency ATP by use of reductive hydrogen, only hydrogen can produce energy. Hydrogen atoms are present in two types: hydrogens marked by (O) bind to oxygen atoms (–OH) and others (*) bind to carbon atoms (–CH). In the cells, these hydrogen atoms are critically different from each other. Cells can use only hydrogens that bind to carbon (reductive hydrogen) for energy production. In 3HB (Fig. 2.1), six reductive hydrogens (–CH) can be used for energy production, although two oxidized hydrogens (–OH) are not used for energy production. Therefore, the capacity of an organic molecule for energy production in cells can be estimated from its chemical structure.

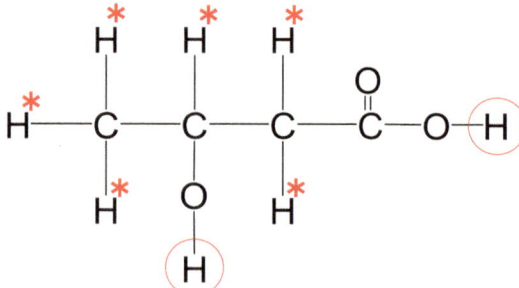

Fig. 2.1 Structure of 3-hydroxybutyrate (3HB), red circles as oxidized hydrogen. 3HB has six reductive hydrogens (*) and two oxidized hydrogens (O). This results that three H_2 molecules can be extracted by oxidation in cells. 3HB is one of the major energy substrates for three reasons; (1) 3HB is metabolized in mitochondria. (2) 3HB has a high energy efficiency. (3) 3HB can pass through the blood-brain barrier

Therefore, why are the hydrogen atoms that bind to oxygen atoms not used for energy production? It should be noted that hydrogen donates an electron, and that oxygen takes an electron (Fig. 2.2). When hydrogen and oxygen bind to each other to form the hydroxyl group (–OH), each atom shares an electron pair with the other. However, this is the "public face." In reality, as an oxygen atom takes an electron, oxygen takes an electron shared by hydrogen. No living organism can take up this electron pair. Protons (H⁺) that do not have an electron cannot be used for energy production.

Although the ketone body has eight hydrogen atoms, living organisms can take the six hydrogen atoms that bind to carbon.

Why can living organisms take electrons from six such hydrogens? Since the force of electronic attraction between carbon and hydrogen is almost equal, they can equally share an electron pair (Fig. 2.3). This is why living organisms can take a

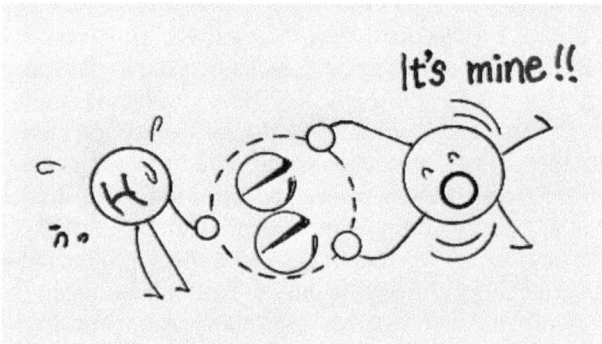

Fig. 2.2 In hydrogen in the OH bond, an electron pair is occupied by oxygen! This is because an oxygen atom strongly absorbs an electron pair. Note that this hydrogen atom cannot be used for energy production, called "oxidized hydrogen"

Fig. 2.3 Hydrogen in the CH bond, an electron pair is shared between hydrogen and carbon! Because carbon and hydrogen atoms absorb an electron pair to a similar extent, they can equally share the pair. Note that this hydrogen atom can be used for energy production, called "reductive hydrogen"

hydrogen atom (H⁺) and an electron from a –CH group, and since hydrogen atoms with an electron are full of energy, they can be used for energy production. Therefore, it is called "reductive hydrogen."

2.3   Energy Currency ATP

Here, I will describe details on energy currency since understanding energy currency is the essential key to grasp the importance of energy problems of the brain, which I will deal with in this book. Since 3HB has six reductive hydrogen atoms, cells can take three hydrogen molecules (H_2). These hydrogens bind to oxygen and produce large amounts of energy, which can be valuable for synthesizing the energy currency ATP.

However, we should delve into a little bit more complex story (oxidative phosphorylation, Fig. 2.4). Reductive hydrogen reacts with oxygen to form water to produce a huge amount of energy, which can be used to phosphorylate ADP to form the intracellular energy currency ATP. The mitochondrion is a specialized organelle for extracting energy from reductive hydrogen and allowing it to react with oxygen to form water. Mitochondria provide the apparatus to couple the oxidation of reductive hydrogen to the phosphorylation of ADP to ATP. This produced ATP is an energy currency of cells. Phosphorylation is the reaction by the dehydration of binding between ADP and phosphate to produce ATP (adenosine 3-phosphate/energy currency). Only mitochondria can do this job, termed oxidative phosphorylation. Oxidative phosphorylation has high energy efficiency. One molecule of hydrogen (H_2) can produce three molecules of ATP. The energy efficiency of this reaction is approximately 40%, similar to the most advanced thermal power plant. We can therefore understand how excellent mitochondria are.

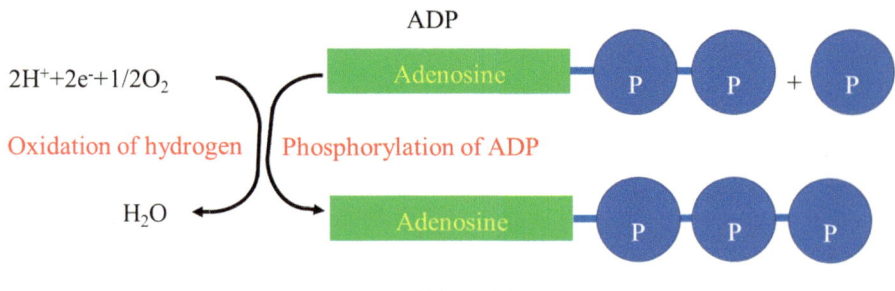

$$2H^+ + 2e^- + 1/2 O_2$$

Oxidation of hydrogen

H_2O

ADP

Adenosine — P — P   +   P

Phosphorylation of ADP

Adenosine — P — P — P

ATP : Energy currency

Fig. 2.4 Roles of oxidative phosphorylation in the energy production in mitochondria. Note that ATP is produced by phosphorylation of ADP. In turn, ATP can give energy to enzymes such as the Na^+ ATP pump by dephosphorylation of ATP to ADP. Since all cells on earth use this energy currency system as an energy source, it can be widely believed that all living organisms originate from the same ancestors

Cells developed the capability of oxidate phosphorylation 2 billion years ago. An aerobic bacteria enters an archaeon and lives inside the cells (symbiosis). After this event, cells could be expanded to form multicellular organisms and, finally, humans with large brains emerged. As I mentioned in the foreword, one of the main goals of this book is the validation of these complex processes from the perspective of the ketone body, a small and simple molecule. The process advanced the history of the earth and finally allowed humans to develop huge brains.

By the way, the human brain has several energy substrates (for the production of ATP). Mitochondria can use fatty acids, ketone bodies, and lactate (lactate converts to pyruvate in the cytosol). In other words, fatty acids, ketone bodies, and pyruvate can directly enter the mitochondria. In the human body, the tissues with the highest production of ATP (the most aerobic tissues) are the heart and the brain. The heart mitochondria demand fatty acids, and the brain mitochondria demand lactate and ketone bodies, as shown in Fig. 2.5.

ATP is hydrolyzed by enzymes to release energy usable for cellular function, as shown in Fig. 2.6. Please remember that most enzymes of the cells just can use energy by the hydrolysis of ATP. Therefore, ATP is used as energy currency in most

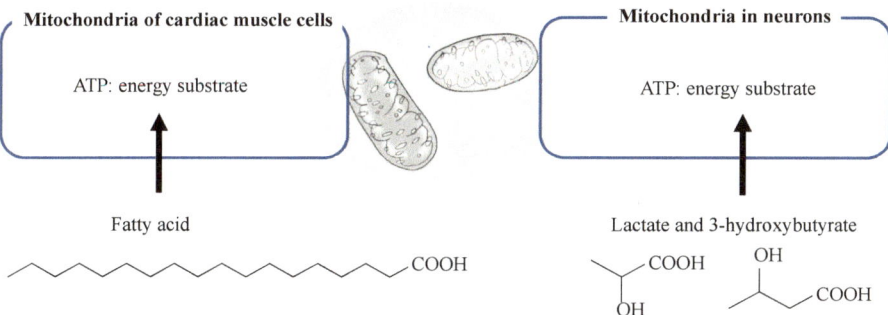

Fig. 2.5 Mitochondria of cardiac muscle cells and neurons. Note that mitochondria produce most of the ATP in the aerobic tissues such as the heart and brain. Although the heart mitochondria preferentially use fatty acid, the brain mitochondria preferentially use lactate and 3HB. Therefore, 3HB can preserve brain function by supplying energy to neurons

Fig. 2.6 ATP functions as an energy substrate. Note that the release of phosphate from ATP produces energy, which can be used in various cellular functions such as the Na⁺ pump

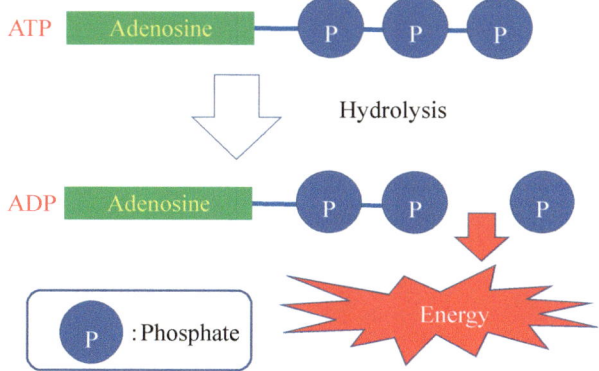

cells on earth, similar to how a key currency, such as US dollars, can be used all over the world. This is why ATP is termed the "energy currency" of energy metabolism. For example, Na^+ pumps extrude Na^+ into the extracellular matrix by using this energy, which is produced by the hydrolysis of ATP, energy currency. Since any living eukaryotic organism cannot survive without ATP, all species on earth share the same energy currency system, suggesting that they evolved from the same ancestors around 4 billion years ago.

2.4   A Lecture on Energy Substrates for Junior High School Students

If I had to deliver a lecture on ketone bodies and energy currency to junior high school students, what would my lecture be? Imagine a teacher who teaches science at some junior high school invites me to give a special lecture. The teacher wants me to educate her students on "energy currency in the biological system." I would go to a nearby stationary store to buy 18 sheets of round cardboard with a diameter of 30 cm, a sheet of magnets with single-sided adhesive, and three colored pens for bold writing. I would be holding them under my arm and go home. After dinner, I try to give an explanation to my family and become frustrated. So, I started working. "C" is written on four cardboard pieces, "H" on eight pieces, and "O" on six pieces. The cardboard was painted with a color pen. Magnet sheets were stuck to the back of cardboard pieces. This takes until midnight, but it is fun. I can feel that the lecture will be very exciting as shown in Fig. 2.7.

Anyway, on the day of giving a lecture, I stood in the face of more than 30 students who were slightly tense and had a somewhat bitter smile. I will put four cardboards of "C" on the blackboard and push them up with my right hand. The

Fig. 2.7 Structure of 3-hydroxybutyrate. Arrows show reductive hydrogen, which is involved in energy production

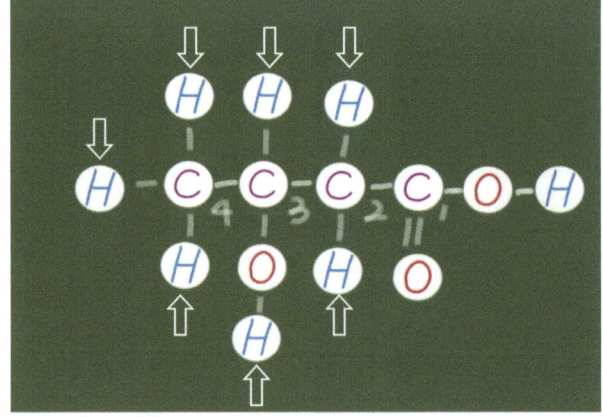

cardboard will be numbered 1, 2, 3, and 4 from right to left. "O" will then be placed on the next right to the first right "C" and "H" will be next to "O." I will silently continue the work. After four "C," eight "H," and three "O" cardboards are positioned across the blackboard, all cardboards are connected by lines with white chalk to complete the chemical structure. Only one line will be a double bond. I'll explain the bond if I have time, but it is good if the students are asking questions.

Well, the class finally gets into the main subject here.

"Look at the blackboard!"

I will take away six cardboard "H" marked with white arrows from the blackboard and dip them into the water bath on the table as well as the prepared three cardboard "O." I roughly flip the shirt's arm and slowly tear off the cardboard in the water. My tension is resolved; I will grab a piece of wet paper, lift it, and say happily. "Energy is released here." The students in the front row get a little bit wet and grimacing momentarily but are not worried about it. Next, I will do this, and I knead a piece of paper with the palm of my hand to make one ball.

"Enzymes in the body make energy like this!"

"When energy is urgently required,"—I will unravel the rolled paper balls into pieces and sprinkle water from the water bath—"This is hydrolysis." At the same time, I sneak my hand into my pants pocket and put something in my hand.

"Enzymes use energy as humans use currency."

I will take a 100-yen coin from unraveled wet papers and say, "You see!"

Although not all students may have fun in my lecture, I will enjoy myself.

2.5 Resting Potential and Action Potential

The next topic is the issue specific to the brain how neurons can evoke action potentials. Neuronal transduction is possible by evoking action potentials. However, if you are familiar with electrophysiology, you can omit this section. The membrane potential of silent neurons is around -80 mV to -90 mV. This is known as the "resting potential." When neurons keep resting potentials, they do not transfer information to other neurons. Then, the ion contents of extracellular and intracellular spaces are different from each other. Although Na^+ concentration is higher in the extracellular space and lower in the intracellular space, K^+ is higher in the intracellular space and lower in the extracellular space, as shown in Fig. 2.8. These features of ion content will not drift because of several neuronal spikes. However, it may be a different story when several hundreds or thousands of action potentials are continuously occurring within seconds. Then, ion contents can change gradually. The big problem is an increase in Na^+ inside the cells. To extrude Na^+ into the extracellular space, cells are installed with a Na^+ pump. Usually, neurons consume

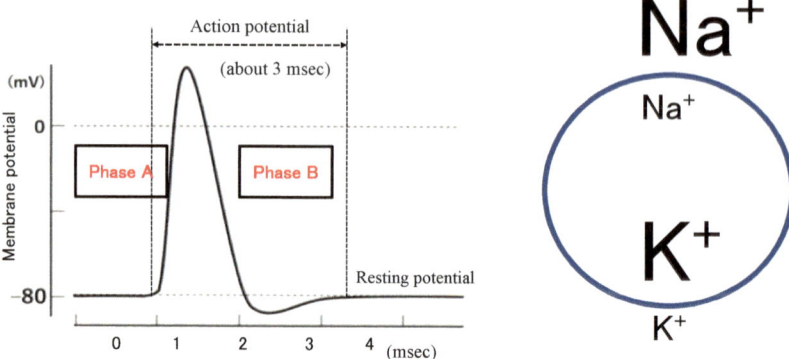

Fig. 2.8 Action potential and ion content in neurons. Action potential has two components, phase A (inward Na⁺ current) and phase B (outward K⁺ current). The resting potential (−80 mV to −90 mV) of neurons is much deeper than other cells (−30 mV to −50 mV). In addition, action potentials are very sharp, and they complete within 3–4 ms, suggesting that neurons are making full efforts by using huge energy to have the readiness to evoke action potentials at any time. The Na⁺ pump gives neurons the readiness to evoke action potentials.

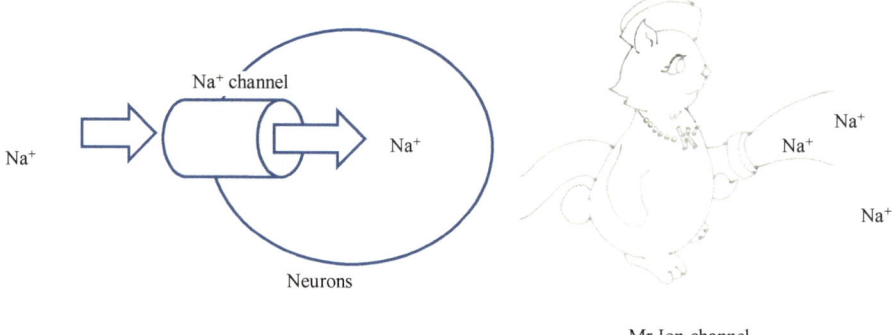

Fig. 2.9 Rapid increase by inward Na⁺ current in membrane potential during phase A. In this phase, membrane potential increases by this inward Na⁺ current. When excitation of the neuronal cell membrane begins, the membrane potential rises sharply. This rise is due to the influx of Na⁺ into the cell. Then, the topic shifts to phase B

a huge amount of energy (around half of energy) for this pump and regulate ion contents within a very restricted range.

To explain what occurs during an action potential, this can be considered to be divided into two phases: the potential is sharply rising in phase A and sharply dropping in phase B. Let us start with phase A (Figs. 2.9 and 2.10).

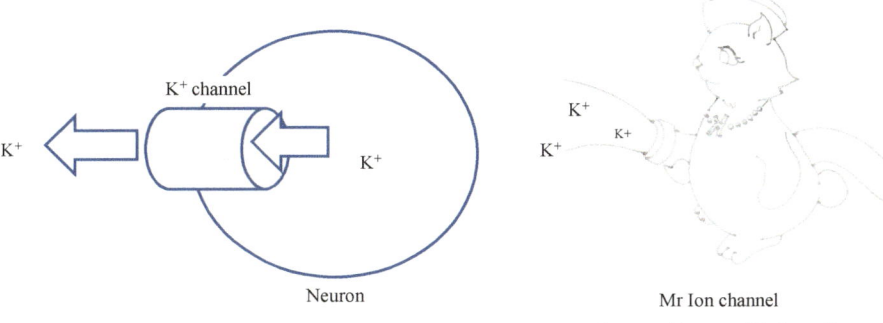

K⁺ channel

K⁺

Neuron

Mr Ion channel

I am working as a K⁺ channel!

Fig. 2.10 Rapid decrease by outward K⁺ current in membrane potential during phase B. In this phase, membrane potential decreases by this outward K⁺ current. Just after the peak is reached, the action potential drops sharply. This decrease is due to the opening of the K⁺ channel. K⁺ flows quickly out into the extracellular space. Through this mechanism, the cell membrane returns to resting potential. This all occurs in a single action potential. All these events are completed within 3–4 ms (a millisecond is one-thousandth of a second!). Neurons are capable of causing 200–300 action potentials within 1 s

2.6   Na⁺ Pump

2.6.1   Pyramidal Neurons with a Big Appetite

Owing to differences in body size, the energy demands of pyramidal neurons may be ten times as large as those of granular neurons. Since 70% of neuronal energy comes from the lactate shuttle supplied by astrocytes, it should be noted how astrocytes perform an essential job.

The reason why pyramidal neurons have a high-energy demand is that they must have an enlarged body size because of their very long axons. (A small cell cannot support a long axon as a dog cannot have an elephant's nose.) The other reason is that they are always ready to evoke action potentials. In addition, they may suddenly accept too many and too frequent action potentials.

Evoking action potentials itself does not require energy use. High energy use is due to readiness to evoke action potentials at any time. This thoughtful act has a large energy requirement. Neurons need to continuously export ions to the extracellular space. Unless the ion concentration is reduced to much lower than that in the extracellular space, neurons cannot attract ions into the intracellular space, that is, neurons cannot evoke action potentials. Neurons do not perform their correct roles and get stuck. In order not to get stuck, neurons must always clean up their actions [2].

The management of sodium ions (Na⁺) is of particular importance. Every time an action potential is generated, it flows into the cell and remains. The Na⁺ pump is the system to deal with this ion, which accounts for about half of the energy consumption of neurons. Na$^+$ certainly accounts for a considerable proportion of energy usage of neurons. It is an amazing fact that neurons can clean the body using only half their energy. Since the Na$^+$ pump is on the cell membrane, the energy substrate (ATP) just below the membrane is used preferentially. Thus, as it is the ATP that produces ATP just below the membrane, almost all of the ATP produced here is consumed by the Na$^+$ pump. As mentioned previously, 30% of energy demand is filled by glycolysis. The ATP used for the Na$^+$ pump is supplied mainly by glycolysis and some fraction of the ATP produced in mitochondria may be used (Fig. 2.11) [3].

The Na$^+$ pump is also known as Na$^+$/K$^+$ ATPase (Fig. 2.1). This means that the substance in the Na$^+$ pump is an enzyme that hydrolyzes ATP. All cells, by using this enzyme, extrude Na$^+$ and take up K$^+$. By using this system, they generate a Na$^+$ concentration gradient across the membrane and can induce action potentials. Consider the following analogy: All types of substances are exchanged between Japan and foreign countries. The Na$^+$ pump produces energy fuel for airplanes that are used for exchange. Since airplanes cannot fly even a millimeter without fuel, this exchange is a very important job [3].

The produced ATP by glycolysis is most consumed by the Na$^+$ pump on the cellular membrane. Continuous driving of the Na$^+$ pump needs a large amount of energy. About 50% of the energy produced in neurons is consumed by this system. Speaking directly, all of the energy produced by glycolysis is spent on the Na$^+$ pump. But the pump requires neurons more energy (20%). This is why pyramidal neurons may often face energy shortages.

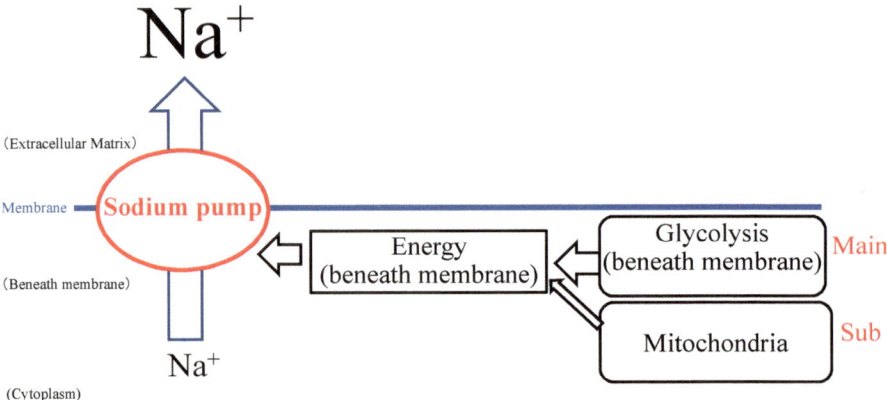

Fig. 2.11 Sodium pump: major energy consumption of neurons [3]. One of the special abilities of neurons is readiness to evoke action potentials at any time. To pursue this special task, neurons spend over 50% of their energy on the Na$^+$ pump. The Na$^+$ pump cleans up Na$^+$ ion from inside cells and absorbs K$^+$ ion from extracellular space by using energy (ATP). Although all neurons are trying hard to meet this demand, pyramidal neurons may often face energy problems when blood glucose levels are drifting. Since glycolysis produces only 30% of total energy, even if all of the energy is consumed by the Na$^+$ pump, the demand is not filled and 20% of energy is supplied from mitochondria

2.6.2   Na⁺ Pump as a Cell Membrane Gatekeeper

In summary, signal transduction is the most important job. Electrical signals mediate signal transduction. The Na^+ pump has a key role. The Na^+ pump is on the cell membrane and hydrolyzes the energy substrate ATP. During this reaction, three Na^+ are extruded and two K^+ are automatically incorporated into cells. The job itself is very simple. All that is occurring is the manual turning of the handle of a window wheel. By doing this, cells exchange inside and outside ions and regulate ion concentrations.

The Na^+ pump is a workaholic gatekeeper to allow the neuron world to exist. Without a gatekeeper, the neuron world cannot function anymore; thus, this role is very important [3]. (Actually, all of the cells in the body have these gatekeepers. They manage every corner of the body.)

As shown in Fig. 2.11, the Na^+ pump is supplied with ATP mainly by glycolysis just beneath the cell membrane. In other words, Na^+ pump and glycolysis are closely connected in terms of energy consumer and energy producer.

2.7   Rescue of Neurons from Overwork

2.7.1   Weak Points of Glycolysis

As I stated previously, 30% of energy demand is filled by glucose. Glycolysis produces the energy substrate ATP by the use of this glucose. Glycolysis has the advantage of supplying energy quickly, but the disadvantage of not providing a sustained energy supply. In addition, it has the serious disadvantage of energy efficiency. The glycolysis can generate just two molecules of ATP per glucose, whereas mitochondria can produce 38 molecules.

The substrate of glycolysis is glucose. Neurons have GLUT3, which allows glucose to pass through the cell membrane. Through this, neurons can receive glucose from astrocytes and provide this for glycolysis just below the membrane. Neurons themselves perform this metabolism. Because the capacity for glycolysis is low and because neurons cannot save glucose, astrocytes must continue to supply glucose without failing. This is why neurons have a big appetite.

In addition, although it is almost a shame to mention, glycolysis has another disadvantage. It induces acidification in the microenvironment just beneath the cell membrane, causing an increase in proton concentrations inside cells. To prevent acidification, neurons must expel protons (H^+) to the extracellular space. This process is automatically performed by influx of Na^+. When Na^+ ingresses, the "timid" protons will escape in response. However, neurons have further work to extrude Na^+ during this job (I will cry; Oh no! I have to keep turning the steering wheel without rest to stay calm. I cannot afford to be calm.) [4].

Neurons need to be supplied energy one after another to clean up the body by using the Na$^+$ pump. Neurons have retained inside the body many small energy-generating units that supply energy to the Na$^+$ pump, which continues the muscle's work of turning the wheel. Neurons have not been occupied by various ions, including Na$^+$ inside cells.

2.7.2 K$^+$ATP Channel Is a Safety Device That Rescues Neurons from Overwork

The small energy-generating units are mitochondria. Pyramidal neurons keep a large number of mitochondria inside the cells. To keep the mitochondria working well, neurons are supplied lactate as energy fuel for mitochondria from astrocytes through the lactate shuttle system.

Neurons require a large amount of energy, even under normal conditions, but they can be in even more difficult situations. (Please imagine that you are working without sleep because of too much work at the end of the year and are completely exhausted.) They may have repeated evoking action potentials for several minutes. Since one action potential lasts 3–4 ms, tens of thousands of action potentials can be repeated during that time. This never happens in daily life.

Repeated action potentials over several minutes are certainly considered "overwork" for a neuron. It is the K$^+$ATP channel that rescues neurons from overwork. (Oh! What a surprise!) Many K$^+$ATP channels exist in neuron membranes. The K$^+$ATP channel activation initiated by the ATP decrease can rescue neurons from overwork. (The K$^+$ATP channel is like a supervisor who says to you, "You should go home." You are grateful to your boss who kindly accepts extra work when he realizes that you are exhausted due to overwork.) [5].

However, some people have difficulty because their K$^+$ATP channels do not work well. In patients with epilepsy, the K$^+$ATP channel is not so sensitive to ATP. When ATP decreases inside neurons, the KATP channel must be activated. However, their K$^+$ATP channel is not activated, even with the ATP decrease.

Thus, it is difficult to rescue them from overwork. This topic of "*refractory epilepsy*" will be discussed in Chap. 11 in which ketone body may be discussed as a lifesaver for patients of refractory epilepsy. Thus, epilepsy is one of the most serious energy problems in the brain.

(Your supervisor goes home first, and you are alone, working overtime. Even when ATP decreases, you are not allowed to have a rest. Then, you repeatedly make mistakes.) This is why you need a helper. (That's it.) As you may notice, excellent helpers are ketone bodies in the brain.

Under these circumstances, it can be said that the energy problem of the brain appears intensively in pyramidal neurons. Pyramidal neurons must be deeply cared for to have a good relationship with a large brain over a long time. There is no choice but to spend the rest of your life with your remaining cells. Let us consider the brain from the perspective of energy. How can we keep our neurons safe throughout our lifetime?

Attention 2.1!

I would like to induce myself. I am the system that can divide glucose into two lactate molecules entering astrocytes through GLUT1. Around ten types of enzymes beneath the membrane place their "hands" together and get the job done like an assembly line. The division of glucose into lactate produces ATP (the cell's energy currency). Well, I am amazing!

[Attention 2.1 quiz]

Who am I?

[Attention 1.4 answer]

Betz's giant pyramidal neurons

I am installed under cell membrane.

Attention 2.2!

To fully understand the Na^+ pump, we have to go back more than 4 billion years.

All of the surface of the ancient earth was covered by sea, in which K^+ concentrations were high and Na^+ concentrations were low. Cells emerged in such an environment. Continents emerged as mountains got higher. Rivers emerged from heavy rain and carved the continent. Ca^{2+} and Na^+ flowed into the sea. Therefore, Na^+ entered the cells. In response to this crisis (a high Na^+ concentration is highly toxic to cells), cells began to pump out Na^+ by using the energy substrate ATP. Cells have continued to use this system. No matter how much time passes, basically nothing has changed. I wish I could help even a little, but I am busy every day. I have many things to do for neurons, including providing energy and recovering wasted fuel.

[Attention 2.2 quiz]

Who am I?

[Attention 2.1 answer]

Glycolysis

I am supplying lactate to neurons.

Attention 2.3!

Today's topic is mitochondria, our best friend.

Photosynthetic bacteria have been producing oxygen for 3 billion years and oxygen concentrations have been gradually rising in the ocean. Many bacteria were seriously troubled by the increase in oxygen in the ocean. Some bacteria began to use hydrogen to detoxify the oxygen and developed the ability to produce the energy substrate ATP to be more active. These bacteria entered another and lived inside the cells, focusing on energy production. They can get what they need from the host and can do what they like. In addition, they can proliferate inside the cells and are happy. This is why mitochondria live symbiotically in cells. I like mitochondria because they are good workers. By the way, mitochondria can put on some extra fat, produce and destroy other energy substrates, and ketone bodies, and obtain huge amounts of currency (ATP). Mitochondria have an outstanding business sense.

I am responsible for K^+ ion flow by opening or closing the doors so that action potentials are not continuous anymore. I am a talented businesswoman and am entrusted to perform environmental management of the office.

[Attention 2.3 quiz]

Who am I?

[Attention 2.2 answer]

Astrocyte

I am a safety device of pyramidal neurons.

References

1. Sandu C, Burloiu CM, Barca DG, Magureanu SA, Craiu DC. Ketogenic diet in patients with GLUT1 deficiency syndrome. Maedica (Bucur). 2019;14(2):93–7.
2. Cunnane SC, Trushina E, Morland C, Prigione A, Casadesus G, Andrews ZB, Beal MF, Bergersen LH, Brinton RD, de la Monte S, Eckert A, Harvey J, Jeggo R, Jhamandas JH, Kann O, la Cour CM, Martin WF, Mithieux G, Moreira PI, Murphy MP, Nave KA, Nuriel T, Oliet SHR, Saudou F, Mattson MP, Swerdlow RH, Millan MJ. Brain energy rescue: an emerging therapeutic concept for neurodegenerative disorders of aging. Nat Rev Drug Discov. 2020;19(9):609–33.
3. Bystriansky JS, Kaplan JH. Sodium pump localization in epithelia. J Bioenerg Biomembr. 2007;39(5–6):373–8.
4. Futai M, Sun-Wada GH, Wada Y, Matsumoto N, Nakanishi-Matsui M. Vacuolar-type ATPase: a proton pump to lysosomal trafficking. Proc Jpn Acad Ser B Phys Biol Sci. 2019;95(6):261–77.
5. Sada N, Lee S, Katsu T, Otsuki T, Inoue T. Epilepsy treatment. Targeting LDH enzymes with a stiripentol analog to treat epilepsy. Science. 2015;347(6228):1362–7.

Chapter 3
Energy Demand of the Human Baby Brain

Abstract Have you ever considered the reason why I am? We know that the substance that allows us to be ourselves is only one. The brain is the substance that is considered the reason. The brain is always considering something. The brain is thinking about what I am eating at dinner. The brain is thinking about how it will be tomorrow. My brain may itself be the reason why I am. Recent studies show that the brain which is always thinking of something is not the privilege of *Homo sapiens*. Other animals are always thinking of something, too. However, a large number of *Homo sapiens* is sharing the same concept although other animals cannot do this. The human brain is consuming a huge amount of energy to share the same concept. As the world will be short in energy if the world depends only on oil as an energy source, the brain may be short in energy if the brain uses only glucose as an energy source. If we wish to solve the energy problem of the brain, we would be able to solve many problems of daily life. Please imagine the days when you can use as much electricity as you like. Although we are working hard to shift to a sustainable society by generating and using reproducible energy, the author thinks that we have to try seriously to engage in the brain energy problems as world energy problems because this is much more closely linked with human health problems. Although the introduction has become longer, let's look into the brain energy problem in terms of its energy substrates, glucose, and ketone bodies.

3.1 Hybrid System

When I was a college student in the 1980s, biochemistry was focusing on energy metabolism. The back cover of textbooks at that time usually had a metabolism map. Glucose was certainly at the center of the map. The metabolism map originated from the Embden-Meyerhof pathway (Glycolysis) which Meyerhof had proposed in the early twentieth century. People of the same generation understood that glucose is located at the center of energy metabolism.

We thought that we understood the human energy metabolism at the center of glucose. However, the human body may be working a little differently. This is the

© The Author(s), under exclusive license to Springer Nature
Switzerland AG 2024
T. Satoh, *Hybrid-Powered Brain*, https://doi.org/10.1007/978-3-031-54150-6_3

central concept of this book. At least, we had thought that we could understand energy metabolism by seeing living creatures at the center of glucose. However, as far as brain energy metabolism is concerned, we can understand energy metabolism only by arranging glucose and ketone bodies located on the two centers side by side.

In addition, when the human brain rapidly expands during late pregnancy, the center of energy metabolism is not glucose but ketone bodies. The total energy of the brain of a human fetus during late pregnancy depends on ketone bodies by 60% according to an old research paper [1]. After birth, the contribution of ketone bodies is rapidly replaced by that of glucose. At least, up to a year after birth, the brain is mainly driven by glucose.

After birth, the brain is generally driven mainly by glucose. One has carbohydrate-based meals three times a day and he has the brain work mainly by glucose during his lifetime. However, we must stop and think. This is because the hybrid is much more efficiently driving the brain than glucose alone. In addition, the brain has a long lifespan (the brain is much more sustained) by a hybrid system. I would like to take a closer look at why a hybrid-powered brain is much more sustainable than a brain dependent on glucose. It will take some time, but please keep in touch until you are satisfied.

3.1.1 The Hybrid System Can Preserve Brain Functions

Homo sapiens has two systems of brain energy. Glucose is from carbohydrates and ketone bodies are from fat. The two systems are not against each other and can cooperate. Specifically speaking, ketone bodies compensate for glucose.

It is very easy to start up the hybrid system. All you have to do is to stop the meals of excessive carbohydrates. You don't have to worry about anything, and you can do it today without much effort. If you have two or three cups of steamed rice because you like white rice, please reduce it to just a cup and increase the amount of side dishes. You must extend the interval between meals and have time to feel small hunger in daily life within a reasonable range.

As mentioned above, the brain is much more efficiently driven by the hybrid system than by glucose alone. Hybrids are much more useful, especially when dealing with dementia. Many papers reported this. Many reports show that cognitive functions are significantly recovered, and neuronal death is suppressed when ketone bodies are orally administered to humans. The hybrid system of glucose and ketone bodies is the most effective method to preserve cognitive functions at high levels. This is based on scientific reason. This is why many researchers focus on the brain hybrid all over the world and several clinical tests are going on. Moreover, there is no need to forcibly supply ketone bodies from the outside, and you can make these in your own body in your daily life. Since ketone bodies can be produced from your fat, this energy system is a highly sustainable system that is popular now. You have to avoid excessive carbohydrate food and your liver keeps synthesizing ketone bodies. All you have to do maybe appreciation for your liver. (It's an analogy of a relay game once again, but you should avoid the situation where there are no players on the day of the big match and no players are waiting for you.) When glucose is in shortage in the case of low blood glucose during drifting blood glucose, we have to avoid the situation where ketone bodies, a powerful reserve player, are absent. In daily life, we have to prepare reserve players (ketone bodies) and we should have the reserve player trained to be ready for the match (situations such as low blood glucose) [1].

3.2   Glucose and 3-Hydroxybutyrate

Here, the chemical comparison between glucose and ketone bodies is mentioned step by step. Please refer to this if you wish to investigate details. Although glucose categorized into carbohydrates has six carbon atoms, 3-hydroxybutyrate (3HB) categorized into short-chain fatty acid or organic acid has four carbon atoms (Fig. 3.1).

Both glucose and 3HB have six reductive hydrogens in their chemical structure (Table 3.1). Cells generate intracellular energy substrate ATP by oxidation of this reductive hydrogen to water (oxidative phosphorylation). Although 3HB has four carbons, 3HB has the same number of reductive hydrogens as glucose. Briefly speaking, the number of reductive hydrogens per carbon atom of glucose and 3HB are 1.5 and 1.0, respectively. 3HB has more hydrogen energy per carbon than glucose does by 50%.

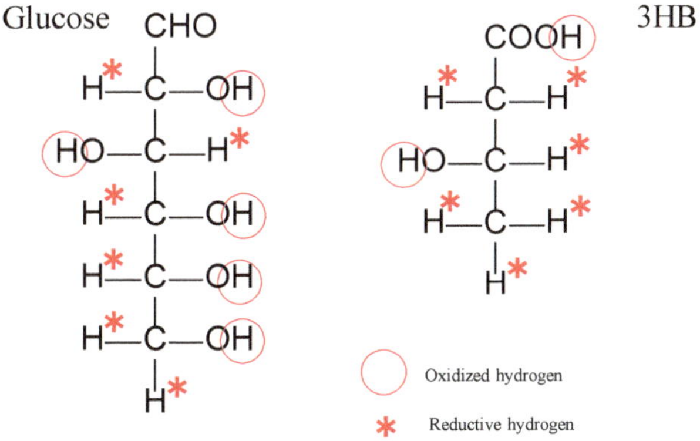

Fig. 3.1 Comparison between glucose and 3HB. Note that they both have six reductive hydrogen, which can be used for energy production although they differ in the number of carbon. They produce the same amount of energy but 3HB is more efficient in terms of energy production per carbon atom

Table 3.1 Chemical comparison between glucose and 3HB

	Glucose	3HB
Reductive hydrogen (=A)	6	6
Number of carbon (B)	6	4
Energy efficiency(=A/B)	1	1.5
The first carbon	–CHO	–COOH
Glycation	Yes	No

3HB has 50% more energy per carbon atom (50% higher energy efficiency) than glucose

Another standpoint of view of comparison is the mode of existence of the first carbon. Although the first carbon is an aldehyde (–CHO) group in glucose, it is a carboxylic acid (–COOH). Since formalin which has an aldehyde (–CHO) group is used for the fixation of biological samples, it easily reacts with amino groups of proteins. Due to these chemical properties, glucose easily degenerates protein structures. This is called "glycation." Through these chemical properties, glucose may degenerate blood vessels and enhance aging. In contrast, 3HB which has a carboxylic acid as the first carbon never induces glycation. In other words, 3HB has no toxic effects on blood vessels although glucose enhances aging by glycation. However, glucose and 3HB share an essential advantage. Both are so highly hydrophilic and so easily dissolved in the blood that they can easily target the brain through systematic circulation. For example, although fatty acid has two reductive hydrogen per carbon and high energy efficiency, it needs to bind to specific proteins due to high hydrophobicity. Thus, fatty acid is not easy to use. Glucose and 3HB are much easier to use than fatty acids. They do not need to bind to proteins because they can themselves dissolve into the blood and freely move to the tissues.

3.3 Human Brain Driven by the Hybrid-Powered Brain

3.3.1 The Adult Human Brain Consumes 23% of Total Energy

This book has the following three goals:

1. To keep the brain at high performance.
2. To have high levels of cognitive functions.
3. To delay brain aging as much as possible.

This book presents these results as a single stacked box. The key to opening the stacked box is to understand the brain's energy system. Once you understand the system, the box will keep open and never close.

First, let's look into the human brain as a whole.

Homo sapiens have a huge brain (1500 g/70 kg) in comparison with other animals. The weight ratio is over 2%. As a side note, the ratio of a chimpanzee is 0.9% (400 g/45 kg), who has high intelligence. *Homo sapiens* certainly has by far the biggest brain out of animals. You may not have a clear image of the weight ratio, but, surprisingly, we look into the energy consumption ratio. The human brain consumes 23% of the total energy of the body. Energy consumes about a quarter, even though it weighs only one-fifteenth. In addition, the brain is on the top of the body. The brain is always lifted and supplied with oxygen and nutrients. (You may imagine the office on the top of the Tokyo Skytree, the highest tower in Tokyo, where thousands of people are working.)

The adult human brain is dependent on glucose as a main energy source although it can use ketone bodies as a backup system. Glucose is certainly an easy-to-use energy substrate. However, there is a big problem. That is the human body stores glycogen (a glucose reserve) just for a half-day use. The amount is by far from that of a single-day use. This is why we must feel hungry in the evening if breakfast and lunch are omitted. Glycogen (a glucose reserve) will begin to decrease by not eating anything a half day. Quantitatively speaking, glycogen (4 kcal/1 g) is stored in the liver and skeletal muscle about 400 g (total 1600 kcal), accounting just for 64% of the energy consumed by a male (2500 kcal). The brain may be injured by energy shortage when humans are exposed to severe starvation. In severe cases, you may fall into a deep coma. However, *Homo sapiens* can stand starvation for at least a month. The people who have experienced fasting for several days know how well fasting affects the brain. The brain is activated, and the emotion is positive, far from deteriorating brain function by not eating for several days. Why does this happen? [2].

3.3.2 Shift to Ketone Energy by Fasting

This is the power of ketone bodies. This is the physiological basis of the concept that the brain should be driven by the hybrid of glucose and ketone bodies. That is why activation of the ketone body shuttle (see below) by extending intervals of a meal as much as possible is by far favorable to the brain and body. Glucose is produced by the hydrolysis of carbohydrates, and ketone bodies are produced by the hydrolysis of neutral fat. Adult humans save about 20 kg of fat as subcutaneous fat or visceral fat in the whole body. This fat corresponds to the energy volume of 72-day use of an adult male (the required energy is 2500 kcal/day). In addition, when he is exposed to starvation, his metabolic system will be adjusted to it and his energy demand will decrease. He can survive over 2 months under complete starvation. However, I do not recommend you have starvation for a long time. Please do not try it [2].

If one cannot synthesize ketone bodies, what will happen to him? It is reported that there are patients with a genetic disease who have deleted the 3-Hydroxy-3-MethylGlutaryl-CoA Synthase 2 (HMGCS2). HMGCS2 is an enzyme which mainly involved in ketone body synthesis. Only eight cases of the disease were reported. They cannot synthesize any ketone bodies. They have a risk of coma when he has not eaten for a half of day. He may have a lot of troubles in their daily life. Every day they are forced to eat something. Scientifically speaking, it is notable that the symptoms are concentrated on the brain. This is physiological evidence that ketone bodies are necessary for the brain [3, 4].

These facts indicate that a smooth shift of energy from glucose to ketone bodies is highly critical for keeping the brain safe in daily life. This is the same story as the smooth shift of gasoline engines and battery motors is the key to the development of high-spec hybrid car engines. The key person of TOYOTA who developed a hybrid car made this comment. As TOYOTA has held the present position by developing this system for the first time, the human brain has held the ecological status by developing the system of smooth shift between glucose and ketone bodies.

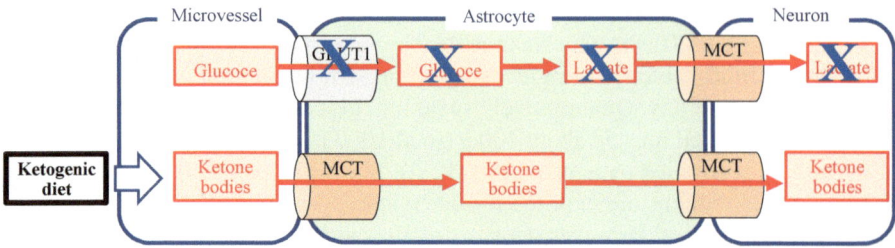

Fig. 3.2 Lactate shuttle does not work in GLUT1 deficiency syndrome [5]. Ketone bodies can support brain function without lactate shuttle. Note that a ketogenic diet can supply ketone bodies to the GLUT1 deficient brain and allow it to work normally

This is not an exaggeration. Please consider the severe situations *Homo sapiens* had faced during these 200 thousand years of their history. Since *Homo sapiens* had stood continuous starvation, one of the most required physiological features is the endurance against starvation. This may be the hybrid-powered brain that allows them to preserve their mental activities under these severe situations. As will be mentioned below, you can understand what an epoch-making innovation the human brain energy system is when you see the process of brain expansion from fetus to baby.

On the contrary, what will happen if the lactate shuttle is not working?

Clinical reports on this case have been numerously accumulating. This case is called "GLUT1 deficiency" as shown in Fig. 3.2. GLUT1 is a transporter of glucose expressed in neurons and astrocytes. The symptoms derived from the brain such as convulsions and impaired consciousness appear. Since astrocytes surrounding microvessels cannot take up glucose due to GLUT1 deficiency, neurons are not supplied with lactate. That is, the lactate shuttle is not working here. Since neurons are supplied 70% of energy from the lactate shuttle, it is natural that pyramidal neurons that demand a huge energy source should become severe energy shortage. This happens in the patient's brain. Since neurons of the patients with GLUT1 deficiency are not supplied lactate from astrocytes, they are suffering from various brain symptoms. However, instead of lactate, the brain normally works by supplying ketone bodies to neurons. The patients can live normal daily lives by continuously having a ketogenic diet. Ketone bodies can compensate the brain energy when the glucose supply is not working well [5].

3.4   Monocarboxylic Acid Transporter (MCT)

3.4.1   The Powerful System Preserves Neuronal Functions

The reason why the roles of ketone bodies remain hard to see is that ketone bodies are prepared as a backup system for the main glucose engine in the adult brain. Glucose is an energy source used for the lactate shuttle which accounts for 70% of the total energy of neurons. The shuttle system is developed as a very advanced and refined communication system between neurons and astrocytes of the brain. Astrocytes and neurons are so closely attached that lactate can directly move from astrocytes to neurons through each transporter protein. This is why most lactate is transferred to neurons without loss. The point is that lactate does not spread out into the cerebrospinal fluid.

Monocarboxylic acid transporter (MCT) connected with astrocytes and neurons is dedicated to a pathway of organic acid and short fatty acid of carbon number 3 or 4 with carboxylic acid at the terminal carbon. MCT is explained in the *"Attention 3.3"* at the end of this chapter. As mentioned above, MCT is initially studied as a pathway for lactate. However, although lactate can go through MCT, this is not only for lactate. Recent studies concluded that MCT is not only for lactate but also for ketone bodies, the energy metabolites which I will talk about in this book. Astrocytes synthesize ketone bodies from fatty acids. It is quite natural that the synthesized ketone bodies should be translocated from astrocytes to neurons [6]. Mitochondria of astrocytes must synthesize ketone bodies even in the adult brain as well as the fetal brain [7]. Unless ketone bodies, the final metabolites of astrocytes, are not translocated into neurons, they must be spread out into cerebrospinal fluid. However, ketone bodies cannot be significantly detected, strongly indicating that synthesized ketone bodies in astrocytes are consumed in neurons. (When all the people waiting in a long line in front of the theater are gone, it's normal to think that almost everyone has gone into the theater hall.)

3.5   Lactate and 3-Hydroxybutyrate

Here, the chemical comparison between lactate and a ketone body (3-hydroxybutyrate, the major compound of ketone bodies) is mentioned step by step. You may be surprised to know how similar chemical structures they have! Let's start by comparing the chemical structure of lactate and 3-hydroxybutyrate (3HB). As shown in Fig. 3.3, 3HB and lactate have similar structures like twins. They share the carboxylic acid of the first carbon and hydroxy group (–OH) in the structures. This is a notable indication in terms of energy metabolism as shown in Fig. 3.3.

Next, please look at Table 3.2. Although 3HB and lactate have similar features to twins, 3HB is 15% (=1.5/1.3) more efficient in energy production. In addition, 3HB can directly enter mitochondria and be metabolized by the use of oxygen.

The brain utilizes energy substrates as shuttle systems between astrocytes and neurons by the following conditions:

(1) High energy efficiency
(2) Existence of a transporter system

Lactate and 3HB share both conditions. In addition, astrocytes can produce lactate and 3HB for neurons. The point is that both lactate and 3HB can pass through the MCT system between astrocytes and neurons, suggesting that both compounds produced in astrocytes are supplied to neurons through a specific transporter.

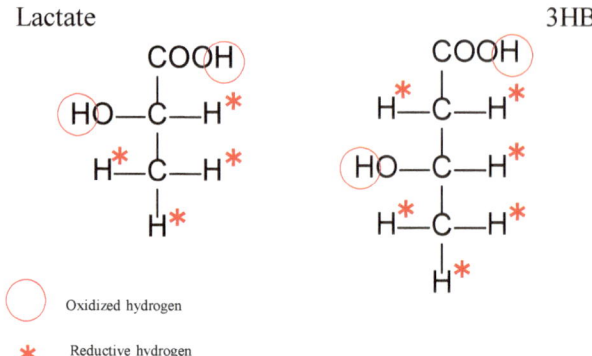

Fig. 3.3 Chemical structures of lactate and 3-hydroxybutyrate (3HB). 3HB and lactate have similar structures. The first carbon is carboxylic acid. The second carbon of lactate and the third of 3HB is the hydroxyl group. However, they critically differ in biological action. Although lactate cannot enter mitochondria, 3HB can directly enter mitochondria. 3HB can recover brain function by supplying energy to neuronal mitochondria

Table 3.2 Chemical comparison between 3HB and lactate

	3HB	Lactate
Reductive hydrogen (=A)	6	4
Number of carbon (B)	4	3
Energy efficiency(=A/B)	1.5	1.3
The first carbon	–COOH	–COOH
Glycation	No	No

3HB has 30% more energy per carbon atom (30% higher energy efficiency) than lactate

Well, what is a different point between lactate and 3HB?

This is the permeability of the mitochondrial membrane. Although both lactate and 3HB enter neurons through MCT and become energy substrate for mitochondrial metabolism, lactate cannot directly enter mitochondria and need to be metabolized to pyruvate in the cytosol of neurons. In contrast, 3HB can directly enter mitochondria and does not need to be metabolized. Thus, 3HB always has an immediate effect.

Lactate is categorized into organic acid and has three carbon atoms. 3HB is categorized into short-chain fatty acid or organic acid and has four carbon atoms.

In contrast, lactate and 3HB have four and six reductive hydrogens, respectively, in their chemical structure. Cells generate intracellular energy substrate ATP by oxidation of this reductive hydrogen to water (oxidative phosphorylation). Briefly speaking, the number of reductive hydrogens per carbon atom of lactate and 3HB are 1.5 and 1.3, respectively. 3HB has more hydrogen energy per carbon than lactate does by 15%. This is a notable indication in terms of energy metabolism. Simply by watching chemical formulas, we can understand that the energy efficiency of ketone bodies is a little (15%) bigger than lactate.

3.6   The Ketone Body Shuttle

Well, I would like to introduce you to a new system here a powerful backup system for the lactate shuttle. Honestly speaking, I finally got here. Since I wish to talk about the system, I have been hitting the keyboards of my computer. This chapter is my feeling. Please slowly read this chapter. A new system is the "ketone body shuttle." The most important key to being with the big brain for a long time is to activate this system continuously. This name will come out many times from now on [8].

Lactate shuttle has been studied by many researchers and is almost regarded as an established theory for neuroscientists. Since lactate and 3HB translocate from astrocytes to neurons through the MCT system, the ketone body shuttle is working as well as the lactate shuttle. Please remember that 3HB and lactate have similar chemical structures to twins, suggesting that 3HB and lactate can pass through MCT. Researchers reported that the "ketone body shuttle" is working in the brain. Manuel Guzman, a professor at the University of Madrid in Spain, focused on the presence and named as "ketone body shuttle" [8]. Ketone body shuttle is popular among neuroscientists who have studied astrocytes, but this is not so well-known among neuroscientists. (Exactly speaking, not many people even know the lactate shuttle.) Although I think that the ketone body shuttle exists in the human brain, so excellent energy system has yet to gain citizenship among scientists around the world. You may be wondering what some hurdles are here. Briefly speaking, direct evidence needs to directly measure the distribution of labeled ketone bodies. There is a heavy hurdle to measuring a labeled compound in a living brain.

The key to understanding energy substrate-mediated communication between astrocytes and neurons is to admit the physiological significance of the ketone body shuttle. Otherwise, it may be forever impossible to understand the brain's energy system. It is no exaggeration to say that we live a normal life with a risk of coma even when blood glucose levels are up and down. As mentioned above, an interesting experiment was performed at Keio University by use of rat embryonic neurons and glial culture. Astrocytes do not synthesize ketone bodies when glucose is stably existing. Whereas they start to synthesize ketone bodies when glucose is decreasing. At least, astrocytes can synthesize ketone bodies from fatty acid when glucose concentrations are decreasing. That is, when glucose is in shortage, astrocytes synthesize ketone bodies to respond to energy shortage. A ketone body shuttle has been installed as a supporting system for neurons to eliminate energy shortage as soon as possible. This is required for astrocytes to help neurons in an emergency [6].

3.6.1   Ketone Bodies for Cognitive Functions

This system has been developed specifically in the brain. One of the big advantages of this system is to use the MCT for lactate. Thanks to this valuable system, the ketone bodies synthesized in astrocytes can rapidly target mitochondria in neurons and be used as an energy source for the production of energy substrate ATP. The relay player's underhand pass, which I mentioned when explaining the lactate shuttle, is also used to deliver the baton called ketone bodies [3, 4].

It is notable that the progression of dementia, where neuronal functions are slowly degenerated, can be delayed by activation of the ketone body shuttle. Many studies report that cognitive functions are recovered by eating a ketone body-rich diet. These reports show the effectiveness of the hybrid power of the brain derived from glucose and ketone bodies. The key to being free from dementia is the activation of the hybrid engine [3, 4].

Ketone body shuttle is for now the hypothesis shared by researchers on astrocytes, including me. However, as shown in Fig. 3.4, it is possible to keep the brain healthy for a long time by introducing the ketone body shuttle. The key to coping with the big brain forever is to activate the ketone body shuttle to some extent [8].

Driving the brain by glucose alone is driving a car with a normal gasoline engine. If you carefully drive a car, fuel economy may be not bad. However, the fuel economy of hybrid engines of gasoline and motor is much better. In this case, the roles of ketone bodies in the brain are the same as those of the motor system in the hybrid car. The main engine must certainly be glucose but using a motor can reduce the load on the engine. Similar to hybrid cars, the roles of ketone bodies are big and critical when blood glucose is up and down. It may determine the survival of the brain [3, 4] (Fig. 3.5).

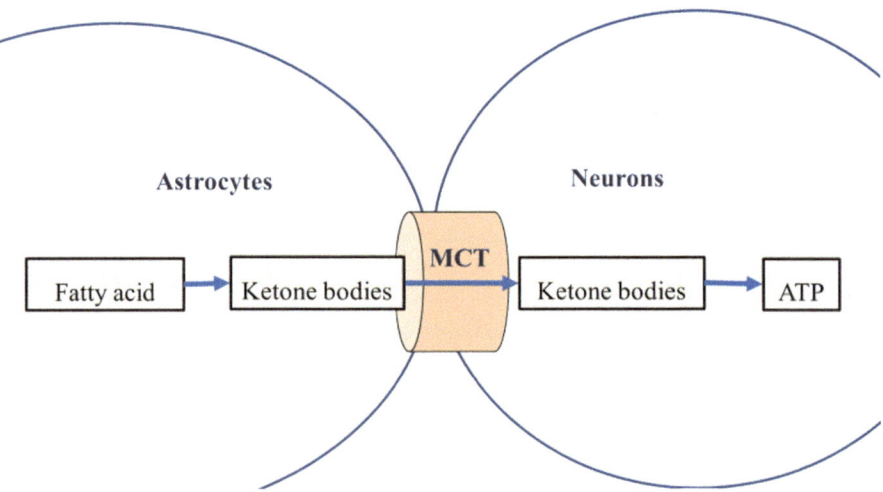

Fig. 3.4 Ketone body shuttle [8]. Specific transporter systems for ketone bodies are working between astrocytes and neurons in the human brain. This shuttle is a critical system to help neurons in the case of energy shortage by drifting blood glucose levels

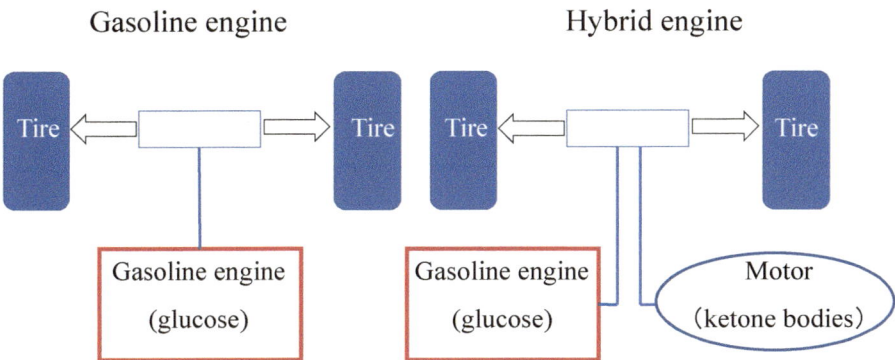

Fig. 3.5 Pyramidal neurons driven by the hybrid engine of glucose and ketone bodies. The brain driven by glucose and ketone bodies is like a hybrid engine of the car, which should have high energy efficiency. In addition, this hybrid system can extend brain longevity and prevent a decline in cognitive function

3.6.2 The Ketone Body Shuttle Is an Essential Subsystem

Although the ketone body shuttle is just a sub-engine, actually it will be a big problem if this is deleted. Pyramidal neurons may be suppressed by energy shortage and become asphyxia due to energy source deficiency when blood glucose is decreasing. Astrocytes cannot rapidly synthesize ketone bodies due to the inhibition of the enzyme responsible for ketone body synthesis by insulin spikes. Since insulin spikes prevent the transducer from the gasoline engine (glucose) to the motor (ketone bodies), the ketone body shuttle cannot work quickly to respond to neuronal crises due to energy shortage [8].

When the energy supply is in shortage, neuronal functions are suppressed, and the brain may become unstable. The brain cannot afford time. This is the same situation as a gasoline engine stops due to a shortage of gasoline. Neurons are often exposed to energy shortage when blood glucose is drifting up and down in daily life. The brain may often have a serious risk of coma. Therefore, the ketone body shuttle is an essential part of the brain's energy to stabilize neuronal functions by eliminating energy shortage.

I hope you can understand the significance of the ketone body shuttle. Although ketone bodies are not visible, it is essential existence to the brain. I cannot wait for the detailed research on the ketone body shuttle. Here we end this section. Thank you for being with us. (In the huge office at the top of Tokyo Skytree, a huge amount of energy is moving around in the cable under the floor. The brain energy substrate has such an image. Ketone body shuttle is like a backup IT system installed for a huge number of computers to work for 24 h.)

Now, let's get back to the beginning. The big advantage of this book is that focuses on the ketone body metabolism of the brain and apply this concept to the eating habit of daily life. By focusing on ketone bodies, we can easily change our daily life drastically. It needs a very small change of eating habits, but it can create a big beneficial effect on the brain. Thus, I call it "small ketogenic."

Attention 3.1!

It's the first time to see you.

I am serving a Hatamaoto (a high-ranked samurai, who has the right to see the Shogun directly). I came here to inform you about glycogen. Glycogen is a polymer of glucose saved in the liver and skeletal muscle. When adrenaline is released, glycogen is hydrolyzed back to glucose. I guess that glycogen is like a Ninja. When accidentally seeing the enemy, running away, and getting excited, glycogen suddenly appears and is hydrolyzed and disappears. He must be a Ninja. However, he often takes care of me, but not directly. Because I can't turn this mechanism without some energy. Hah, hah! My job is to adjust the ion concentrations of neurons. If I stop the job, neurons cannot evoke action potentials. I can't cut corners at all.

[Attention 3.1 quiz]

Who am I?

[Attention 2.3 answer]

K⁺ATP channel

I am consuming a half of energy of neuros!

Attention 3.2!

Hallow!

Because the author mentioned a chimpanzee, I will talk a little about the evolution from monkey to human. I will talk about the reason why *Homo sapiens* has a big brain.

Point 1: Humans started upright bipedal walking. For this reason, they can use their hands freely and start making things. They began to consider something to make something.

Point 2: They stood upright. They could make various voices by expanding their throats. Thus, they can speak words.

Point 3: Chimpanzees fight other animals and kill them by bite. When humans appeared, they stopped killing other animals by bite. Muscles that move the jaw become thinner, and thanks to that, the ventricles can be expanded. You can create something new by stopping another thing. However, I do not quit this job. I know anybody has trouble if I quit the job. I am a follower of Na^+ pump and I am entrusted with the task of creating the path of ions. It's amazing!

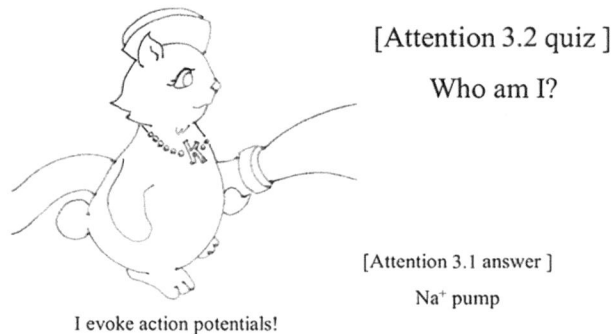

[Attention 3.2 quiz]

Who am I?

[Attention 3.1 answer]

Na^+ pump

I evoke action potentials!

Attention 3.3!

My full name is a monocarboxylic acid transporter. My name originates from – COOH (carboxylic acid). These are short organic acids.

Acetate as C (Carbon) number 2

Lactate as C (Carbon) number 3

Ketone body as C (Carbon) number 4

All of them have a single carboxylic acid. Transporter means "A path for such carboxylic acids." Well, you may understand us a little. Anyway, when I run, it's really fast. I will continue to run.

Bow-wow! Bow-wow!

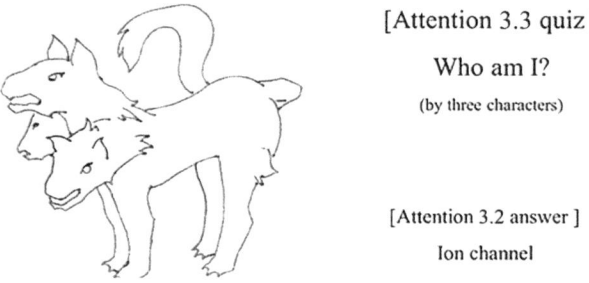

[Attention 3.3 quiz]

Who am I?

(by three characters)

[Attention 3.2 answer]

Ion channel

I am a path of cell membrane for ketone body!

References

1. Adam PA, Räihä N, Rahiala EL, Kekomäki M. Oxidation of glucose and D-B-OH-butyrate by the early human fetal brain. Acta Paediatr Scand. 1975;64(1):17–24.
2. Geisler CE, Ghimire S, Bogan RL, Renquist BJ. Role of ketone signaling in the hepatic response to fasting. Am J Physiol Gastrointest Liver Physiol. 2019;316(5):G623–31.
3. Jensen NJ, Wodschow HZ, Nilsson M, Rungby J. Effects of ketone bodies on brain metabolism and function in neurodegenerative diseases. Int J Mol Sci. 2020;21(22):8767.
4. VanItallie TB. Biomarkers, ketone bodies, and the prevention of Alzheimer's disease. Metabolism. 2015;64(3 Suppl 1):S51–7.
5. Sandu C, Burloiu CM, Barca DG, Magureanu SA, Craiu DC. Ketogenic diet in patients with GLUT1 deficiency syndrome. Maedica (Bucur). 2019;14(2):93–7.
6. Takahashi S, Iizumi T, Mashima K, Abe T, Suzuki N. Roles and regulation of ketogenesis in cultured astroglia and neurons under hypoxia and hypoglycemia. ASN Neuro. 2014;6(5):1759091414550997.
7. Bélanger M, Allaman I, Magistretti PJ. Brain energy metabolism: focus on astrocyte-neuron metabolic cooperation. Cell Metab. 2011;14(6):724–38.
8. Guzmán M, Blázquez C. Is there an astrocyte-neuron ketone body shuttle? Trends Endocrinol Metab. 2001;12(4):169–73.

Chapter 4
Energy Demand of the Human Fetal Brain

Abstract The baby's brain consumes most of the produced energy. Even at 1 year of age, an infant's brain uses 53% of the energy in the body. Therefore, I will explain in a detailed manner how this large energy demand is supplied. Why does a newborn human's brain consume 74% of the energy produced in the whole body? The human infant's brain receives huge information, especially visual information, just after birth. It is a situation called excessive input. Especially, apes and birds of prey with binocular vision have large visual cortexes in the neocortex and special functions of constructing three-dimensional visions from large visual information on the plane coming from the eyeballs. This tendency is especially noticeable in humans. A baby's brain is simultaneously processing huge visual information and performing synaptic rearrangement. (Behind that cute look, it is surprising that the baby's brain is steadily performing the work like a supercomputer.)

4.1 The Newborn Brain Consumes 74% of the Energy

Please look at Table 4.1. Although the adult brain just occupies 23% of energy (2% in weight), the fetal brain accounts for 74% (11% in weight).

The newborn brain consumes most of the energy produced in the whole body. Even in a year baby uses 53%, over half of energy is consumed in the brain. Next, I will explain to you in a detailed manner how this huge energy demand is supplied. Why does a newborn brain consume 74% of the energy produced in the whole body?

Table 4.1 Percentage of brain weight and energy consumption in humans (%) [1]

	Brain weight	Brain energy consumption
Newborn	11	74
6-month-old	12	64
1-year-old	10	53
5 years old	7	44
10 years old	4	34
15 years old	3	27
Adult	2	23

T. Satoh, *Hybrid-Powered Brain*, https://doi.org/10.1007/978-3-031-54150-6_4

For these purposes, the brain needs a lot of energy. As mentioned in Chap. 1, the brain mainly comprises two kinds of cells, neurons and glia. Since birth glial cells continue to grow until after birth, they consume some energy.

Although neurons stop growing in late pregnancy, they have to do the following two important jobs after birth:

1. Synapse recombination
2. Information processing from the outside world

It can be easily predicted that both of these two jobs need huge energy. This is why 74% of energy is spent in the brain (Table 4.1). From the standpoint of view of energy metabolism, humans may have reconstructed their whole body for energy supply to their brain.

Although the speed of the increasing number of neurons slowed down during late pregnancy, glial cells including astrocytes were continuously growing. The brain is bigger and bigger every day during the late pregnancy. This expansion is due to an increasing number of glia but not neurons. The brain engages in synaptic rearrangement, and synapse formation and elimination. The late pregnancy of humans is highly distinctive from other animals because most of the energy is consumed by the brain. This phase concentrates on the expansion of the brain in terms of energy consumption. In addition, the main energy substrate is not glucose but ketone bodies. The ketone bodies are produced mainly by the placenta.

Well, how is the huge energy source supplied to the brain? The next central issue is the mechanism of delivery of energy substrate to the brain.

4.2   Energy Substrates of Newborn Rat Brain

First, let's see the rat brain. Table 4.2 compares the velocities of energy substrate uptake into the brain between newborn and adult rats. The velocity of ketone body uptake into the brain decreased tenfold as rats developed from newborns to adults. Whereas that of glucose uptake increased to 2.5 folds. These results show that the ketone body is dominant in the embryonic brain. During the late pregnancy to the newborn stage, the rat brain produces energy by use of a ketone body-dominant system. However, the brain becomes glucose-dominant as it develops. In addition, up to the newborn state, rat embryo has HMGCS2, which produces ketone bodies, in most tissues. The concentrations of ketone bodies are well correlated with the levels of expression of HMGCS2 responsible for the synthesis of ketone bodies since the HMGCS2 is one of the slowest steps in a metabolic pathway of ketone

Table 4.2 Maximum rate of uptake of energy substrate in the brain (rat) [2]

	Glucose		Ketone bodies	
Stage	Newborn	Adult	Newborn	Adult
Maximum velocity (mmol/g/min)	0.52	1.22	2.01	0.22

body synthesis from fatty acid in the mitochondria of hepatocytes. However, after the maturation, the rat has expressed HMGCS2 just in the liver and the epithelium of the gut.

The changes in ketone body concentrations confirmed that the ketone body-dominant brain is shifted to the glucose-dominant brain (Table 4.3). The concentrations of ketone bodies are 1.0 mM at 5 days after birth and the high concentrations are maintained for 20 days. However, after 20 days after birth, the concentrations are sharply decreasing. The reduction of the velocity of ketone body uptake is due to the decrease in ketone body concentrations (Table 4.3). This sharp decline of ketone bodies must be caused by a decrease in the HMGCS2 expression except in the liver and the epithelium of the gut.

Several papers reported that newborn rats have high levels of HMGCS2 in many organs and tissues (Table 4.4). Though HMGCS2 is highly expressed in the liver and small intestine (digestive organs derived from endoderm), most tissues converge to very low levels up to the weaning period (28 days old). The only exception is the liver, which lifelong maintains relatively high levels of ketone body synthesis. According to these data, ketone body concentrations are highly maintained up to the weaning period in mammals, suggesting that ketone bodies are the main energy sources not only in the brain but also in the whole body.

Notably, ketone body synthesis is not high in the central nervous system tissues (cerebellum, neocortex, medulla oblongata, and midbrain). Most of the ketone bodies consumed in the rat brain may be supplied from the liver and small intestine. However, the exception is astrocytes in the rat brain, which express high levels of HMGCS2 and have the ability of ketone body synthesis (Table 4.5).

Table 4.3 Blood ketone body concentration in rats [3]

Days after birth	5	10	20	30
Ketone body concentrations (mM)	0.99	0.86	0.75	0.12

Table 4.4 Changes in HMGCS2 mRNA expression level in each organ (rat) [4]

	Lactation period (11 days old)	Weaning period (28 days old)
Liver	4300	500
Heart	40	20
Kidney	520	50
Small intestine	5500	30
Cerebellum	80	10
Neocortex	40	0
Medulla oblongata	50	10
Midbrain	50	0

Table 4.5 Expression of HMGCS2 in rat astroglia cells [4]

	mRNA levels of HMGCS2
Newborn	370
Adult	25

Table 4.6 Uptake of glucose and ketone bodies in the brain of a 28-week-old human fetus [5]

	Substrate uptake rate (μmol/min•kg)
Glucose	17.5 (40%)
Ketone bodies	27.3 (60%)

4.3   Energy Substrates of the Human Fetal Brain

Since the advent of *Homo erectus* (2 million years ago), the human brain has grown rapidly. The human brain has to engage in the physiological problem of how much energy is supplied to the brain, which has high energy demand and is located in the highest position. The next topic is the energy problem of the human brain. The main energy sources of the human fetal brain are ketone bodies, too as discussed here (Table 4.6). Forty-eight years ago, the authors of the paper measured the velocities of glucose and ketone body uptakes of the brain, which was taken out from an aborted human fetus (28-week-old) and cultured for several hours in an organ level. I was surprised to find that such an important work by use of the human fetal brain had been unknown to the world [5]. According to this report, ketone bodies and glucose account for 60% and 40%, respectively of energy, providing direct evidence that the human fetal brain preferentially uses ketone bodies as energy sources. The human brain can rapidly grow because of the energy system of the main ketone bodies and the sub-glucose system. This important physiological finding has been overlooked until now although this paper made this finding public as if the veil had been removed. Thus, I wish to make this finding open to the public again so that ketone bodies are the main energy sources of the human fetal brain (Table 4.6). At least, up to the late pregnancy, the human brain may be driven by the ketone body fuel.

4.4   What Is the Energy Fuel of the Human Fetal Brain?

4.4.1   Ketone Bodies Are the Main Energy of the Human Fetal Brain

Where do ketone bodies come from to supply it to the human fetal brain? It is reported that the concentrations of ketone bodies in the human fetus are significantly high compared with those of adult humans. A human fetus is certainly dependent on ketone bodies as the main energy sources of the brain and whole body, as shown in Table 4.7. In addition, the ketone bodies supplied to the fetus are synthesized in the fetal side of the placenta. The ketone body concentration must be the highest in placental fetal villi. This is because the fetal side of placental fetal villi may be the biggest supplier of ketone bodies to human fetuses, especially to the fetal brain.

Table 4.7 Ketone body concentration in human fetus (mM) [6]

Placental villi	2.235
Cord blood	0.779
Systemic circulation	0.2
Cerebrospinal fluid	1.977
Maternal blood	0.1

Ketone body levels decrease from placental villi (2.235 mM), cord blood (0.779 mM), and fetal blood (0.200 mM). Please be careful about the order of the concentrations. One reasonable interpretation is that the placenta mainly produces ketone bodies and that the synthesized ketone bodies are spread out into the fetal systematic circulation through cord blood. In the human fetus, the most important flow of the ketone bodies is supposed to be the following:

Placental villi → cord blood → fetal systemic circulation (Fig. 4.1).

Although the human brain is supposed to grow rapidly by supplying ketone bodies as an energy source from the placenta, there must be another supplier to the brain. That is because the brain substrate demand is huge and accounts for over 70% of the total energy. As shown in Fig. 4.1, astrocytes largely expressing HMGCS2 also produce ketone bodies to maintain the concentrations at high levels in the brain. Ketone bodies produced in astrocytes are spread out into the cerebrospinal fluid to supply ketone bodies to the neurons because the ketone body shuttle is not fully grown as seen in the human fetal brain.

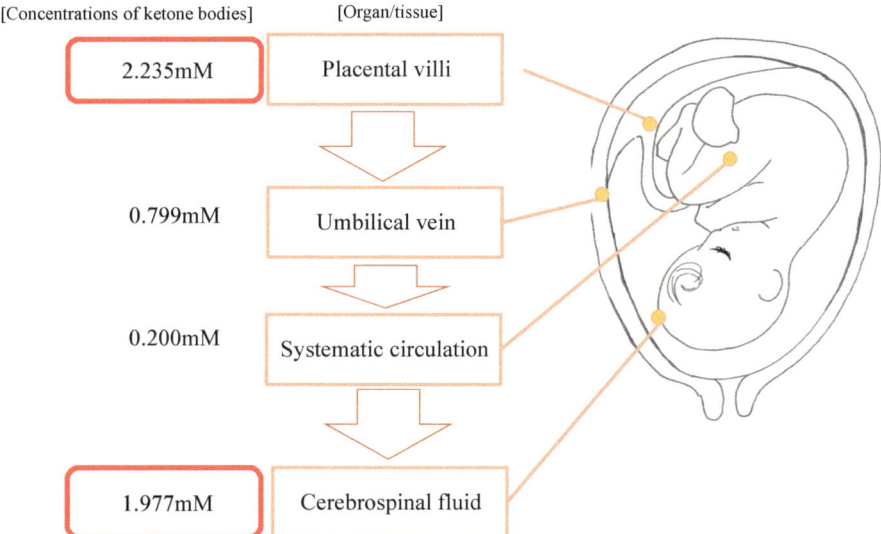

Fig. 4.1 Concentrations of ketone bodies in the human fetus [6]. Ketone body concentrations are gradually decreased from placental villi-umbilical vein-fetal systemic circulation. In contrast, the concentrations are unexpectedly high in the cerebrospinal fluid, suggesting that astrocytes supply ketone bodies to neurons in the fetal brain

4.4.2   Astrocyte Production of Ketone Bodies in the Human Fetal Brain

Notably, the ketone body concentrations of cerebrospinal fluid are much higher (1.977 mM) than those in the systematic circulation (0.200 mM) during late pregnancy. The reasonable estimation is that ketone body concentration in cerebrospinal fluid is much lower than that in blood. However, the ketone bodies are ten times more concentrated in cerebrospinal fluid. The reasonable estimation is that there is a cellular group producing ketone bodies in the cerebrospinal fluid. According to the previous reports, the cellular group must be astrocytes in the brain. There is no published research to demonstrate that human primary fetal astrocytes produce ketone bodies as far as I know. However, according to *in vitro* data by use of rat or mouse astrocytes, the ketone bodies-producing cells must be astrocytes. By these backgrounds, I assume the following flow of ketone bodies in the human fetus:

astrocytes → cerebrospinal fluid → neurons

Ketone body concentrations of the cerebrospinal fluid of adult humans are not increasing. Ketone bodies are directly moving from astrocytes to neurons through the monocarboxylic transporter (MCT) in the adult brain. In contrast, the MCT system may NOT be fully installed in the human fetal brain. For this reason, ketone bodies produced in astrocytes are easily spread out into the cerebrospinal fluid. This is why the ketone body concentrations in the cerebrospinal fluid of the human fetus are much higher than those of adult humans. However, this is not enough. We assumed that there must be another supplier of ketone bodies to the human fetal brain [6].

4.5   The Placenta Is a Main Supplier of Ketone Bodies

4.5.1   Two Pathways of Ketone Body Supply from the Placenta

The placenta is one of the most important suppliers of ketone bodies to human fetuses in late pregnancy. Since most energy substrates are devoted to the development of the brain during late pregnancy, placental production of ketone bodies is dedicated to fetal brain development. We measured the 3HB (the major compound of ketone bodies) concentrations in the amniotic fluid, umbilical vein, umbilical artery, and maternal blood (Fig. 4.2).

The concentrations of ketone bodies in the amniotic fluid were significantly higher than those in the maternal blood. In addition, those in the umbilical vein and artery were also higher than those in the maternal blood [8]. Thus, the concentrations of ketone bodies in all of the pathways to the human fetus from the placenta are high, suggesting that the placenta may serve as a potent producer of ketone bodies in the human fetus. As shown in Fig. 4.3, the placenta has two cell populations

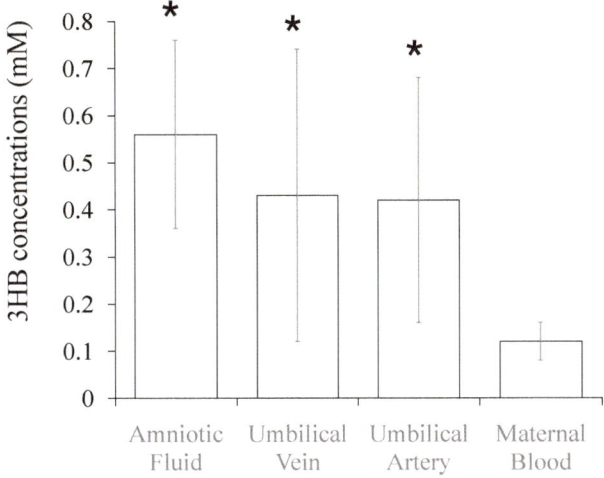

Fig. 4.2 High concentrations of 3HB in the uterus [7]. 3HB concentrations of amniotic fluid, umbilical vein, and umbilical artery are significantly (*) higher than maternal blood. 3HB produced in the placenta is supplied to the human fetus. The 3HB is essential for brain development

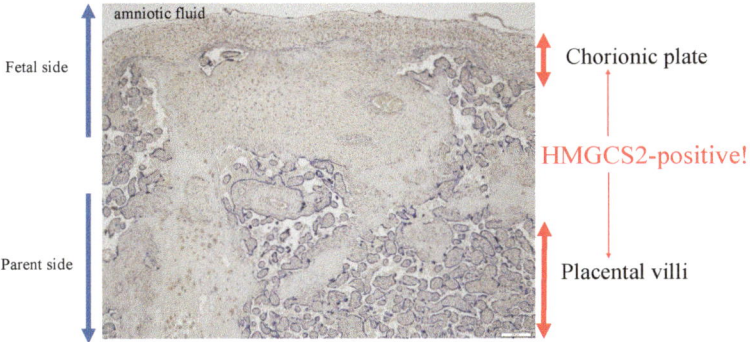

Fig. 4.3 Expression of HMGCS2 (an enzyme for ketone body synthesis) [7]. HMGCS2 expression was examined in the human placenta using immunohistochemistry right after delivery. The human placenta displays two 3HB producers, villi and the chorionic plate to supply 3HB to the umbilical vein and amniotic fluid, respectively

that produce ketone bodies for the human fetus (HMGCS2-positive cells), tropho-blast cells of placental villi and those of chorionic plate. Based on these results, there are potential two pathways of ketone body delivery from the placenta to the human fetus:

1. Chorionic plate → amniotic fluid → small intestine → systemic circulation
2. Placental villi → cord blood → systemic circulation

   The human fetus may be able to be supplied with ketone bodies by the use of the ketone body production system (HMGCS2) of his own liver and small intestine.

The human fetal brain is supposed to get bigger and bigger during the late pregnancy and this development must highly require the supply of ketone bodies from the placenta. Placental supply of ketone bodies also may help the fetus to supply ketone bodies to the brain.

Although ketone bodies are synthesized in the fetal side of the microvilli of the placenta and astrocytes in the brain, most of the glucose is transferred to the fetus from maternal blood through her placenta. Well, how does the brain of a human fetus get ketone bodies? It strikes me that a human fetus is trying hard to analyze the method with its brain and procure ketone bodies. The word "miracle of the brain" comes across my head. The brain of a human fetus keeps the following original pathways of ketone body delivery. Although they were already mentioned, I write them down again because this ketone body transfer is an important step to the brain growing big. The placenta serves as a powerful supplier of ketone bodies to the human fetal brain.

The brain of the human fetus is quietly recombining synapses by the use of ketone bodies as main energy substrates. This is how the brain is growing big by the hybrid of glucose and ketone bodies in late pregnancy. A human fetus is preparing the birth on earth as a single person and as the most evolved multicellular organism (What a reliable system! I feel the beauty of life science in this system?) [7].

4.6 Distinctive Human and Rat Brains

Rats and humans share the basic structure of the brain and have big differences. Here, I explain these. First, there is an anatomical difference. In the human brain, the neocortex is huge and covers the other brain. Many large wrinkles are carved to increase the area as much as possible. In contrast, the rat brain is called a smooth brain without a few wrinkles. The human brain is a mass of fat and soft. Rat brain is less fat and much harder. Although I had horse brain on my hand during the veterinary course at my university (of course, the horse is naturally dead), I had two strong impressions of the brain. One is that the horse brain has a large neocortex and many wrinkles, similar to humans. The other is that it is like a mass of fat and very soft. If you hold it with a little force, it will soon collapse. Human and horse brain has a large fat, whereas rat brain has less fat and few wrinkles. The rat brain is somewhat hard, and the substance is clogged. Human fetuses and rat embryos have the same energy metabolisms by the following two points [7]:

Both brains have ketone bodies as the main energy source.
After birth, ketone bodies are rapidly replaced by glucose.

Furthermore, the human fetal brain has the following specific features of energy metabolism. Ketone body concentrations are reduced to 0.2 mM in the systemic circulation during the late pregnancy of human fetus although rat fetus maintains 1.0 mM ketone bodies. Synthesis of ketone bodies in the liver and small intestine is already low during the late pregnancy of human fetus although rat embryo maintains the high expression of HMGCS2. Rat embryo has considerably high ketone body synthesis in the liver and small intestine.

The basic flow of ketone bodies of the human fetus may come from the placenta through the following two pathways:

1. Placental villi → cord blood → ystemic circulation → fetal brain
2. Chorionic plate → amniotic fluid → small intestine → systemic circulation → fetal brain

Although the liver and small intestine play the role of ketone body synthesis in the rat embryo, the placenta plays the role in the human fetus. This flow of ketone bodies is the main key to the human brain growing big in late pregnancy.

The human fetus has high levels of ketone bodies in the cerebrospinal fluid. This ketone body is another key to the human brain growing big in late pregnancy. By these backgrounds, I propose a novel pathway of ketone bodies supply to the fetus's brain as shown in Fig. 4.4.

Fatty acid supplied by the mother converts to ketone bodies by HMGCS2 of trophoblast cells of the fetal side of the placental villi, which targets the fetal brain. The ketone bodies account for the main source of energy of the growing fetal brain.

As a supplemental device for ketone body production, astrocytes have much HMGCS2 and produce ketone bodies, which are supplied to neurons through cerebrospinal fluid.

By use of ketone bodies supplied by the two pathways, mitochondria of human fetal neurons synthesize the energy substrate ATP, which is used for synaptic recombination of the developing brain [8].

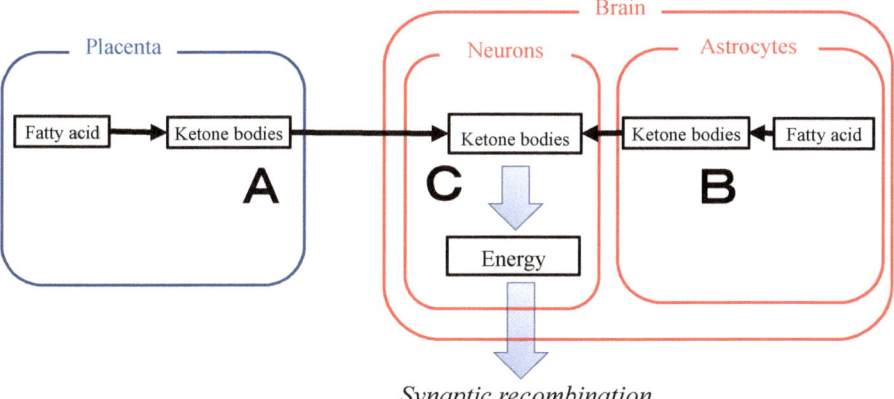

Fig. 4.4 Development of human brain needs for ketone bodies supplied by astrocytes and placenta during late pregnancy [7]. The human fetus has three production facilities of ketone bodies: (A) placenta, (B) liver, and (C) astrocytes. At least, during late pregnancy, the placenta (A) is the biggest supplier of ketone bodies. Therefore, the human placenta is a unique organ by the feature that serves as an effective producer of ketone bodies. This supply must be essential for the expansion of the human fetal brain during late pregnancy

Everything is mysteriously, magically, and steadily performed according to tens of millions of years of history. By the way, both two energy sources, glucose supplied from the mother and ketone bodies synthesized in the fetal side of placental villi and astrocytes contribute to the growing embryonic brain. This can be said a miracle of the brain or a miracle of glucose and ketone bodies.

Attention 4.1!

How do you do?

I am appearing so late because it's hard to get away from the axon. The human brain enthusiastically engages in the recombination of synapses just after birth. This is why layer IV of the neocortex is getting thick. Vision on the plane of the eyes is processed in the neocortex so they can know the distance because it is three-dimensional in your brain. I guess that only humans can see full-colored three-dimensional vision. Well, Troodon could see binocular vision. They had almost the same performance as owls. Troodon is the closest group of theropods. Oh! I talked about unnecessary things. I have to return to axons. I am rolling around axons so tightly that action potentials from neurons should not leak out. My job is highly rewarding.

[Attention 4.1 quiz]

Who am I?

[Attention 3.3 answer]

MCT

I am an insulator of axons!

Attention 4.2!

I explained the digestive system organs derived from the endoderm. The endoderm initially develops as the epithelium of the gastrointestinal tract of many animals. The basic body plan of all metazoan animals is very simple, like a chub. Even an early human fetus has such a form of a simple chub. The chub has three layers from the inside to the outside, endoderm, mesoderm, and ectoderm. The epithelium of the endoderm grows and sinks to the mesoderm and ectoderm and forms the liver, esophagus, stomach, small intestine, and large intestine. I wish to talk more about these topics. But I have to leave here because new action potentials are coming. I am the hub of the neuronal circuits. Granular neurons input information and I deliver far away by collecting the information. Farewell!

[Attention 4.2 quiz]

Who am I?

[Attention 4.1 answer]

Oligodendrocyte

I am bigger than granular neurons!

Attention 4.3!

Long time no see, everyone!

I would like to talk a little about the advent of *Homo erectus* 2 million years ago. Then, humans began to use fire. By using fire, they started cooking. This is why they can often eat animal protein and fat. By this eating habit, humans can synthesize ketone bodies and supply it to the brain. This allows the human brain to grow rapidly. However, this is just a scientific hypothesis. *Homo erectus* may be your (*Homo sapience*) direct ancestor. *Homo sapiens* has a huge brain and is always thinking of many things, for example, the energy supply system of the human brain. That's very amazing. Please use huge energy in your brain. Best regards.

[Attention 4.3 quiz]

Who am I?

[Attention 4.2 answer]

Pyramidal neurons

I produce ketone body!

Attention 4.4!

I am here explaining the phospholipid that is involved in the ketone body synthesis in astrocytes. The cellular membrane is made of phospholipids. An enzyme is coming and digging up fatty acid by a scoop. The enzyme synthesizes ketone bodies by use of fatty acid. By this mechanism, astrocytes have a limited capacity of ketone bodies. After all, it's a sober action. Astrocytes have a hard task. I have a hard task, too. I am always cleaning up every corner of the castle of my lord. I am always keeping the concentrations of Na^+ and K^+ constant. Boo-hoo-hoo!

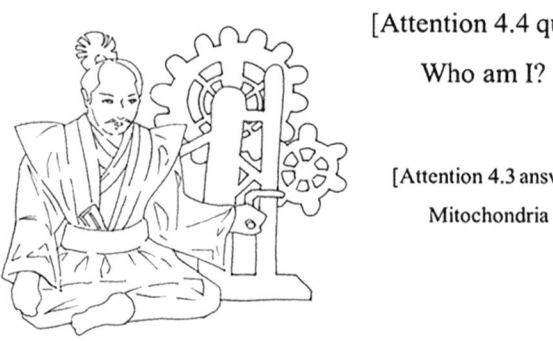

[Attention 4.4 quiz]

Who am I?

[Attention 4.3 answer]

Mitochondria

I keep ion contents constant!

Attention 4.5!

Well, everyone!

Ketone bodies are the main energy sources of the brains of human fetuses and rat embryos. What does this mean? My opinion is that ketone bodies are the main players in the developing brains of human fetuses and rat embryos. In addition, ketone bodies are important subs energy sources of the brain even after birth as well as the brains of human fetuses. Farewell! I am a family doctor of neurons and guiding to heaven in the case of death.

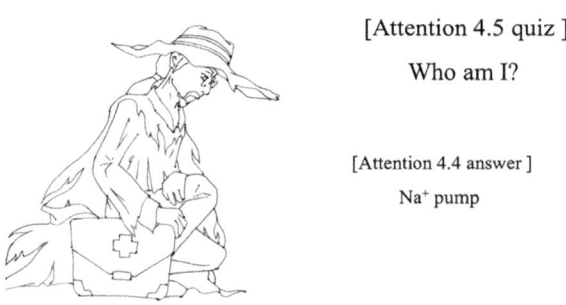

[Attention 4.5 quiz]

Who am I?

[Attention 4.4 answer]

Na^+ pump

Previously, I was regarded as a funeral director !

References

1. Cunnane SC, Crawford MA. Survival of the fattest: fat babies were the key to the evolution of the large human brain. Comp Biochem Physiol A Mol Integr Physiol. 2003;136(1):17–26.
2. Cremer JE. Substrate utilization and brain development. J Cereb Blood Flow Metab. 1982;2(4):394–407.
3. Lockwood EA, Bailey E. The course of ketosis and the activity of key enzymes of ketogenesis and ketone-body utilization during the development of the postnatal rat. Biochem J. 1971;124(1):249–54.
4. Cullingford TE, Dolphin CT, Bhakoo KK, Peuchen S. CoA lyase and detection of the corresponding mRNA and those encoding the remaining enzymes comprising the ketogenic 3-hydroxy-3-methylglutaryl-CoA cycle in the central nervous system of suckling rat. Biochem J. 1998;329:373–81.
5. Adam PA, Räihä N, Rahiala EL, Kekomäki M. Oxidation of glucose and D-B-OH-butyrate by the early human fetal brain. Acta Paediatr Scand. 1975;64(1):17–24.
6. Muneta T, Kawaguchi E, Nagai Y, Matsumoto M, Ebe K, Watanabe H, Bando H. Ketone body elevation in the placenta, umbilical cord, newborn and mother in normal deliver. Glycative Stress Res. 2016;3:133–40.
7. Satoh T, Shibata T, Takata E, Takakura M, Han J, Yamada S. High 3-hydroxybutyrate concentrations in the placenta-produced amniotic fluid in the human uterus. medRxiv 2023.08.09.23293873.
8. Edmond J. Energy metabolism in developing brain cells. Can J Physiol Pharmacol. 1992;70(Suppl):S118–29.

Chapter 5
Supply of Energy Substrate to the Brain

Abstract In this book, I have tried to discriminate the contents that most scientists agree with from those parts of the scientist's share. In this chapter, I will make the points simple; thus, the discriminations may be slightly unclear. Let us start to talk about human adult brain energy metabolism. Neurons depend on astrocytes for a huge energy supply. Neurons receive 30% of their energy from glucose and 70% from lactate. Neurons cannot survive for several minutes without energy supply from astrocytes. The brain has a distinctive energy supply system from other organs, that is, the lactate shuttle and the ketone body shuttle. Neurons are supplied by two energy substrates, lactate, and ketone bodies. Lactate is produced in astrocytes from glucose. "Shuttle" means the lactate transfer system between astrocytes and neurons that makes the transporters so closely located that lactate could move directly from astrocytes to neurons. Recently, the ketone bodies have been working between astrocytes and neurons, known as the "ketone body shuttle."

5.1 Shuttle Systems

5.1.1 Cooperation Between Ketone Bodies and Glucose

The ketone body shuttle and lactate shuttle work between astrocytes and neurons to fill the huge energy demand of neurons. Thus, the chemical comparison between lactate and ketone bodies is an interesting point, too. Of course, the lactate shuttle is an excellent system as long as blood glucose is stable. However, the brain may face serious problems when blood glucose is drifting up and down. The time will appear when the astrocyte cannot procure glucose for driving the lactate shuttle. At this time, pyramidal neurons are exposed to severe energy shortages. For the worse, the brain may become coma. In response to these risks, the brain has a ketone body shuttle as a backup system. Because the human fetus's brain is working mainly by the ketone body shuttle, this backup system is highly effective. Even if the lactate shuttle is not working at all (i.e., GLUT1 deficiency), the ketone body shuttle alone can drive the brain. On the contrary, if the ketone body shuttle is now working (i.e., HMGCS2 deficiency), the lactate shuttle alone stands the brain function.

You can understand that the lactate shuttle and ketone body shuttle are working cooperatively by knowing the cases of GLUT1 deficiency and HMGCS2 deficiency. Briefly speaking, GLUT1 deficiency is a lack of lactate shuttle and HMGCS2 is a lack of ketone body shuttle [1, 2]. In addition, the lactate shuttle and ketone body shuttle can fully complement each other.

Although astrocytes of patients with GLUT1 deficiency cannot uptake glucose from microlevels, they can live a normal life if neurons are supplied with ketone bodies from the liver by eating the habit of the ketogenic diet. Even in this case, the patients must stand on the continuous ketogenic diet. In addition, patients with HMGCS2 deficiency can also live a normal life by eating a carbohydrate-rich diet. These two genetic diseases show that the lactate shuttle and ketone body shuttle can cooperate in the human brain [3, 4].

5.1.2 The Liver Has a Large Capacity for Producing Ketone Bodies

Please keep in mind the following things. Most of the brain is made of fat. A total of 70% of the contents except water are fat. However, over half of fat is cholesterol, which is mainly saved in oligodendrocytes. This fat cannot be used for ketone body synthesis. The amount of fat, which can be used for ketone body synthesis in astrocytes is highly limited.

By the way, how do astrocytes procure fatty acid, which is going to be used for ketone body production? Astrocytes may cut off phospholipids in cellular membranes to procure fatty acid. Otherwise, they store natural fat inside cells to cut off fatty acid from this store. In each case, mitochondria are the key to the production of ketone bodies. Mitochondria converts carbohydrates to neutral fat by use of acetyl CoA synthesized from glucose. However, the amount of production is highly limited. For similar reasons, astrocytes cannot fully use fat, which accounts for 70% of the brain. The amount of fat that can be used for the synthesis of ketone bodies within the brain is highly limited [3, 4].

Well, the topic will return to the synthesis of ketone bodies in astrocytes. Astrocytes can supply neurons with ketone bodies through transporters. However, this is just a backup system for the lactate shuttle. Thus, it can support neuronal functions just for a short time. Fatty acids may cross through the blood-brain barrier, but it is not certain that the amount is sufficient for total brain function [3, 4]. (It's like forgetting your wallet and buying a ticket with the coins left in your pocket. You can't go that far.)

However, ketone bodies can pass through the blood-brain barrier. The liver cells, as well as astrocytes, can synthesize ketone bodies and directly send them out to neurons. (Is it an image that the country quickly and sufficiently delivers relief supplies and human resources to the disaster-stricken area?) By this system, neurons maintain neuronal functions as if they were completely safe and have no trouble. The brain becomes more active by the supply of ketone bodies from the liver [3, 4]. (Returning to the ticket story, it's like going through a ticket gate using a smartphone payment app, even though you've forgotten your wallet. If this is the case, you can even get on a limited express train.)

5.1.3 Excessive Carbohydrate Intake Causes Damage to the Brain

Well, as a bonus of this chapter, let's consider the reason why we must not have excessed carbohydrate meal. The damage induced by excess carbohydrates is mainly due to frequent insulin spikes. Insulin spikes completely suppress HMGCS2, an enzyme responsible for ketone body synthesis. For this reason, ketone bodies will not work when it is required. As a result, the brain is not working well. Briefly speaking:

1. Astrocytes in the brain have a highly effective system (ketone body shuttle) for supplying ketone bodies to neurons exposed to glucose shortage (i.e., when blood glucose is decreased).
2. However, the amount of fat in the brain is limited, that can be used in astrocytes for ketone body production.
3. For this reason, the liver usually sends out ketone bodies to the brain and expands the amount of ketone bodies delivered to the brain to protect neurons in the case of emergency (low blood glucose).
4. However, excessed carbohydrate meals eliminate the ketone body production by astrocytes and the liver.
5. The brain will more frequently not work well.
6. In the long run, one may fall into depression or dementia.

Be careful about your eating habit of you and your family and let's look into this again. I don't want to write something close to a threat like 5 or 6, but I thought that writing this much might bring back healthy days, so please forgive me.

5.2 Glucose Targets to Brain

The first thing I would like you to know is that astrocytes fill all the gaps between neurons as shown in Fig. 5.1. In addition, astrocytes wrap around almost all blood vessels in the brain as shown in Fig. 5.2. Therefore, astrocytes are involved in all of the delivery of energy substrates to neurons as passing points. Any energy source cannot target neurons independently on astrocytes. (Both glucose and ketone bodies can enter the king's room by meeting with a butler named "astrocyte." In addition, the butler is always preparing lactate from glucose and ketone bodies from fat for the master named as "neuron") [5, 6].

Astrocytes → Neurons

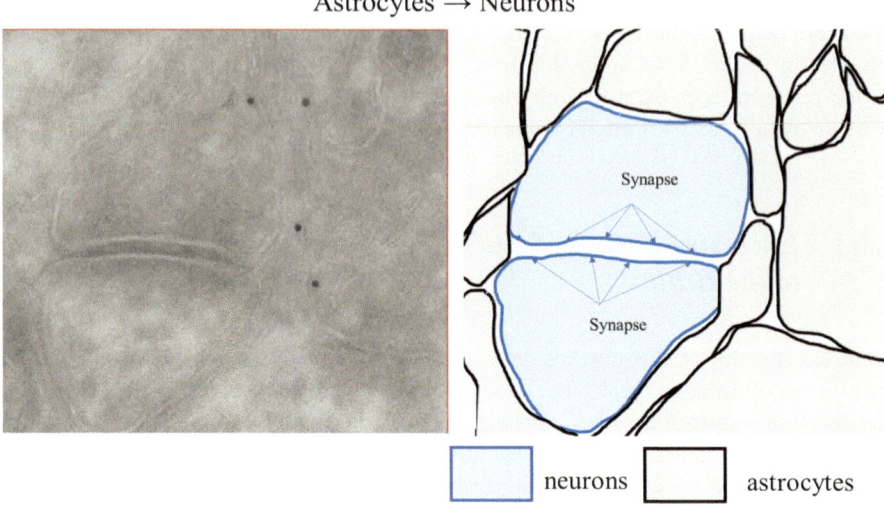

neurons   astrocytes

Fig. 5.1 Astrocytes fill the gap between neurons. Astrocytes are filling all of the gaps in the brain. Therefore, astrocytes may be involved in various functions of the brain including synaptic transmission

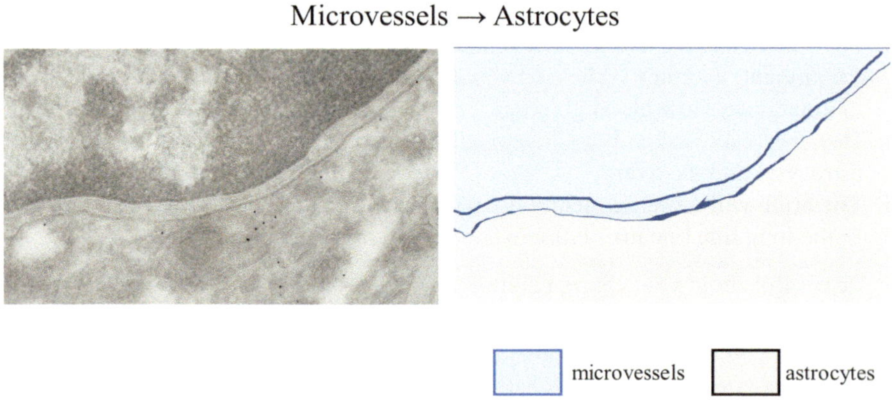

Microvessels → Astrocytes

microvessels   astrocytes

Fig. 5.2 Glucose supply to neurons mediated by astrocytes. All nutrients and oxygen of the microvessels are passing through astrocytes to neurons

Therefore, the transport of energy substrates from cerebral microvessels to neurons has the following sequence:

microvessels→astrocytes→neurons

Well, first let's consider the pathway of glucose targeting to the brain after carbohydrate intake.

[Fig. 5.3: Route 1]

Carbohydrate (meal)→glucose (digestive tract)→(portal vein)→(liver)→(systematic circulation)→(cerebrovascular)→*GLUT1*(micro vessel~astrocyte)→*lactate pro-*

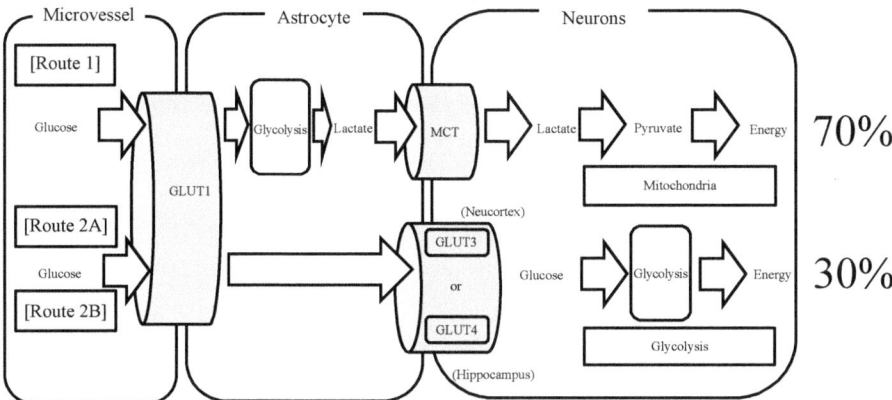

Fig. 5.3 Supply route of glucose to neurons [5, 6]. The glucose in the microvessels has two pathways to neurons. One is lactate shuttle (70%) and the other is glucose itself (30%). The principal energy of neurons is lactate supplied through MCT. Glucose, occupying 30% of total energy, drives the Na⁺ pump beneath the cell membrane to prepare evoking action potentials

duced by glycolysis→MCT (astrocyte~neuron)→lactate→pyruvate→energy production mitochondria

When carbohydrate is eaten, it is rapidly hydrolyzed by enzymes to produce glucose. Glucose is absorbed by epithelial cells of the small intestine and enters into systematic circulation through the portal vein. Glucose is absorbed by foot protrusion of astrocytes, which wrap around cerebrovascular, and coverts glucose to lactate. Then, lactate is transferred to neurons. The lactate converts to pyruvate by an enzyme, which can target mitochondria in neurons and is used for energy production. This pathway supplies 70% of the amount of energy consumed in neurons. This is called as "lactate shuttle."

Glucose and lactate can pass through GLUT1 (a transporter of glucose) and MCT (a transporter for lactate and ketone bodies), respectively. By these two systems, glucose and lactate can directly move over the gap of cells. These compounds are never spreading out into cerebrospinal fluid. They can pass through astrocytes to neurons. (As mentioned above, neurons are so-called gluttons with a very strong appetite. Otherwise, neurons are in a bad mood and can't work, and cause many problems. It can't be helped.)

5.2.1 Insulin Is Essential for Glucose Uptake in the Hippocampus

This system (Route 2A and Route 2B) supplies neurons with glucose for glycolytic production of ATP. Most of the produced ATP beneath the cell membrane is consumed by the Na⁺ pump, which extrudes Na⁺ ions from neurons. In conclusion, glycolysis (an energy producer) and Na⁺ pump (an energy consumer) are closely connected. Briefly speaking, the ATP supplied through this pathway is largely connected with maintaining of ion homeostasis of neurons, which is essential for evoking action potentials [7, 8].

[Fig. 5.3: Route 2A: neocortex]

Glucose (cerebrovascular)→GLUT1(cerebrovascular~astrocyte)→*GLUT3(astrocyt c~neuron)*→glucose (neocortex)→*energy production in glycolysis*

[Fig. 5.3: Route 2B: hippocampus]

Glucose (cerebrovascular)→GLUT1(cerebrovascular~astrocyte)→*GLUT4(astrocyt e~neuron)*→glucose (hippocampus)→energy production in glycolysis

Neuron directly takes up glucose, which accounts for the remaining 30% of energy consumed in neurons. Glucose passing through foot protrusion of astrocytes enters neurons and is used for energy production in glycolysis. Glucose transporters in neurons have two kinds GLUT3 (neocortex) and GLUT4 (hippocampus).

What is different between GLUT3 and GLUT4 although both are glucose transporters? The answer is a requirement of insulin. GLUT3 does not require insulin although GLUT4 requires insulin. Although pyramidal neurons of the neocortex take up glucose independently from insulin, pyramidal neurons of the hippocampus strictly require insulin to take up glucose. For this reason, pyramidal neurons cannot form a short-term memory when neurons are insulin-resistant. I do not recommend you increase insulin spikes. Maybe you have to save insulin spikes in order not to become insulin resistant. Notably, pyramidal neurons of the hippocampus are the most sensitive tissue to insulin resistance.

Most of the ATP produced by glycolysis just beneath the cell membrane is consumed by the Na^+ pump. Thus, the glucose transported through GLUT3 (neocortex) or GLUT4 (hippocampus) is highly connected with the maintenance of the Na^+ gradient produced by the Na^+ pump. In conclusion, there is a close connection between the consumer (Na^+ pump) and producer (glycolysis) beneath the cell membrane of pyramidal neurons.

What is the difference between Route 2A (GLUT3) and Route 2B (GLUT4)? The expression of glucose transporters is critically different between GLUT3 and GLUT4 as shown in Table 5.1. Since pyramidal neurons in the neocortex (GLUT3) absorb glucose independently from the presence of insulin, long-term memories are not affected by diabetes. In contrast, since pyramidal neurons in the hippocampus (GLUT4) take glucose dependently on the presence of insulin, short-term memories are greatly affected by diabetes, suggesting that diabetic patients are apt to catch dementia, too. If insulin is not well working (type 2 diabetes), hippocampal pyramidal neurons cannot uptake glucose. In contrast, cortical pyramidal neurons are not

Table 5.1 GLUT3 and GLUT4

GLUT type	GLUT3	GLUT4
Organ of expression	Neocortex	Hippocampus
Memory	Long-term memory	Short-term memory
Insulin-dependency	No	Yes

affected. This is a critical reason why dementia is initiated by neuronal degeneration of the hippocampus. If you catch type 2 diabetes, you may be likely to catch dementia.

5.3   Ketone Bodies Target to Brain

[Fig. 5.4: Route 3]

Neutral fat (fat tissue)→fatty acid (hydrolyzed in adipocytes)→(systematic circulation)→ketone bodies (the liver)→(systematic circulation)→(cerebrovascu lar)→*MCT* (cerebrovascular~astrocyte)→*MCT* (astrocyte~neuron)→*energy pro-duction in mitochondria*

Ketone bodies increase by stopping excessive carbohydrate meals. Ketone bodies enhance the hydrolysis of fat stored in the fat tissue. Let's consider how ketone bodies target neurons of the brain. Ketone bodies can rescue energy shortage of the brain because of freely passing through the BBB.

Neutral fat abundantly (over 20 kg in adult man) in fat tissue is hydrolyzed by enzymes since insulin spikes are eliminated. For this reason, fatty acids are released into the blood. The hepatic cells take up fatty acids to produce ketone bodies, which are released into systematic circulation and reach cerebrovascular. Finally, astrocytes take up ketone bodies and send them out to neurons.

Since *Route 3* is just a supporting system to glucose against the extreme reduction of carbohydrates, the contribution is not big by normal eating habits. Ketone bodies reach the mitochondria of neurons through astrocytes and are used as an energy source. Ketone bodies can use MCT when they pass through the gap between astrocytes and neurons [9, 10]. The MCT is the same molecule that mediates the lactate shuttle. (This is an image like limited express trains can pass through the tracks on which rapid trains usually run.)

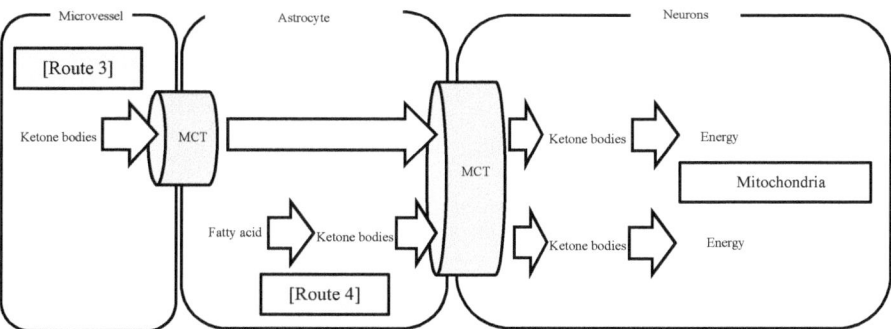

Fig. 5.4 Supply of ketone bodies to neurons [5, 6]. Neurons are supplied with ketone bodies by the two pathways. One is from the liver and the other is from astrocyte. Ketone bodies are directly targeting neuronal mitochondria for energy production

[Fig. 5.4: Route 4]

Neutral fat or phospholipid→fatty acid→ketone bodies (astrocyte)→*MCT* (astrocyte~neuron)→*energy production in mitochondria*

Ketone bodies synthesized from astrocytes translocate into neurons and function as energy substrates. Since astrocyte has limited storage of fat, it can support neuronal functions for a short time. However, it is a highly effective system when blood glucose is unstable. On the contrary, this system is working when blood glucose is decreasing. This is usually stored in an emergency warehouse. This is called as "ketone body shuttle." The products, ketone bodies, are directly delivered from the factory. This is highly reliable in an emergency.

We have overviewed the pathway of energy sources (glucose and ketone bodies) leading to neurons. What do you think of this system? This is a little complicated, isn't this? We deliver energy sources to neurons and allow neurons to work every day by using a pathway or pathways properly depending on the time and the case. It is amazing to consider, sometimes it is OK, how this reaches the brain or whether it is used in glycolysis when you are eating. (I have never considered such a thing when I have a meal.)

Attention 5.1!

I am back again.

I am in trouble because of protons staying here inside neurons. I cannot be against the order of the Na^+ pump. Well, let's talk about fat hydrolysis. Fat is hydrolyzed to produce acetyl Co-A and finally, energy substrate ATP is produced. However, acetyl Co-A is synthesized back to fat in the case of frequent insulin spikes.

Well, let's talk about fat synthesis. When glucose is hydrolyzed to acetyl Co-A through pyruvate, acetyl Co-A is synthesized back to fat or glycogen. But it is not a bad thing because you can use it in the future. For this reason, humans can save the fat for the resistance to starvation. Well, I will be back to my job. I help ions flow in and out. This job is rather exciting. I have a good name for "membrane protein."

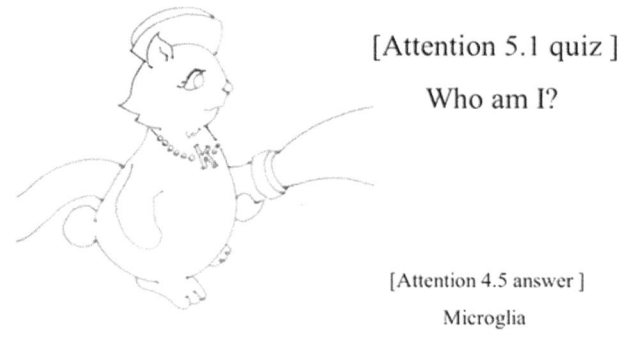

[Attention 5.1 quiz]

Who am I?

[Attention 4.5 answer]
Microglia

I am a path of K⁺ or Na⁺!

Attention 5.2!

I am Mr. Glycolysis. Here I am talking about the relationship between GLUT4 and insulin. GLUT4 is a gate for glucose on the cell membrane, but it needs a key named "insulin." Since pyramidal neurons of the hippocampus have GLUT4, glucose cannot enter the cells unless neurons catch insulin. In addition, since most of all cells have GLUT4, glucose cannot enter the cells when insulin is absent. What I mean is that insulin is so very important hormone. We have to appreciate it. If you overuse insulin, insulin may become inactive. So, you should be careful. Because glucose is entering neurons, I have to be back. See you again!

[Attention 5.2 quiz]
Who am I?

[Attention 5.1 answer]
Ion channel

I have a close relation with Na⁺ pump.

Attention 5.3! (The Last Attention)

I am here to explain Routes 3 and 4. Route 3 prevents energy shortage and recovers the health of neurons. Route 4 plays the role of the rescue team to save lives by providing first aid (ketone bodies) to neurons. When a first aid is emitted, we will deliver the drip in a hurry because it is affecting the survival of neurons. We always serve neurons by teaming up around the neurons in large numbers. We take care of meals as well as maintaining health. We are working hard in all of the gaps between neurons. There are a few gaps where we are not working. This is like the night sky filled with a vast number of stars. The support system is so perfect. Oh! Sir Neuron is calling me. I am sure that he is in an energy shortage. I have to prepare ketone bodies for emergency use. Well, I will leave here. I always pray that your neurons are healthy as long as you live.

[Attention 5.3 quiz]

Who am I?

[Attention 5.2 answer]

Glycolysis

[Attention 5.3 answer]

Astrocyte

I am supplying ketone body to neurons.

References

1. Sandu C, Burloiu CM, Barca DG, Magureanu SA, Craiu DC. Ketogenic diet in patients with GLUT1 deficiency syndrome. Maedica (Bucur). 2019;14(2):93–7.
2. Ago Y, Otsuka H, Sasai H, Abdelkreem E, Nakama M, Aoyama Y, Matsumoto H, Fujiki R, Ohara O, Akiyama K, Fukui K, Watanabe Y, Nakajima Y, Ohnishi H, Ito T, Fukao T. Japanese patients with mitochondrial 3-hydroxy-3-methylglutaryl-CoA synthase deficiency: in vitro functional analysis of five novel HMGCS2 mutations. Exp Ther Med. 2020;20(5):39.
3. Nakamura MT, Yudell BE, Loor JJ. Regulation of energy metabolism by long-chain fatty acids. Prog Lipid Res. 2014;53:124–44.
4. Ferguson BS, Rogatzki MJ, Goodwin ML, Kane DA, Rightmire Z, Gladden LB. Lactate metabolism: historical context, prior misinterpretations, and current understanding. Eur J Appl Physiol. 2018;118(4):691–728.
5. Segarra M, Aburto MR, Acker-Palmer A. Blood-brain barrier dynamics to maintain brain homeostasis. Trends Neurosci. 2021;44(5):393–405.
6. Magistretti PJ, Allaman I. A cellular perspective on brain energy metabolism and functional imaging. Neuron. 2015;86(4):883–901.

7. Szablewski L. Glucose transporters in brain: in health and in Alzheimer's disease. J Alzheimers Dis. 2017;55(4):1307–20.
8. López-Gambero AJ, Martínez F, Salazar K, Cifuentes M, Nualart F. Brain glucose-sensing mechanism and energy homeostasis. Mol Neurobiol. 2019;56(2):769–96.
9. Cunnane SC, Trushina E, Morland C, Prigione A, Casadesus G, Andrews ZB, Beal MF, Bergersen LH, Brinton RD, de la Monte S, Eckert A, Harvey J, Jeggo R, Jhamandas JH, Kann O, la Cour CM, Martin WF, Mithieux G, Moreira PI, Murphy MP, Nave KA, Nuriel T, Oliet SHR, Saudou F, Mattson MP, Swerdlow RH, Millan MJ. Brain energy rescue: an emerging therapeutic concept for neurodegenerative disorders of ageing. Nat Rev Drug Discov. 2020;19(9):609–33.
10. Guzmán M, Blázquez C. Is there an astrocyte-neuron ketone body shuttle? Trends Endocrinol Metab. 2001;12(4):169–73.

Chapter 6
Commitment to a Dementia-Free Society

Abstract Your brain is always thinking of something. Even now your brain is thinking of something. Ten seconds later it is thinking of something and 20 s later it is thinking of something. The brain is always thinking continuously. Note that the brain is consuming a large amount of energy to allow the brain to think continuously. Please imagine that you cannot think continuously and how your daily is. Ten seconds later you may forget what you are thinking of. Specifically speaking, you have experienced having forgotten what you are going to do when you go upstairs. Everybody has ever done this and of course I have. However, you must have serious trouble in your daily life if you always forget what you were thinking of 10 s before. The human brain distinguishes between the past, the present, and the future and allows you to think continuously. As you get older, the brain does not allow you to think continuously, and your thinking becomes non-continuous. When this gets worse, you have lost the past and the future and you are in the present. You may be diagnosed with dementia. The brain being continuous is highly related to the energy problem that I stated in Chap. 2. This is because the human brain is consuming a large amount of energy to allow you to think continuously. The energy problem of the brain is highly related to the issue of how people are living in the super-aging society, which the USA, Europe, and Japan are now facing. No one likes to get dementia, and anybody wishes his or her family not to get dementia. This is why we are discussing with each other and can get to a better way to fact dementia without fear by letting the public know about the brain energy problem. In this chapter, we pursue a possible way to protect the brain, especially the hippocampus which manages short-term memory, and the neocortex which controls long-term memory.

6.1 Short- and Long-Term Memories

6.1.1 Fragment and Connected Information

"What did you eat a week ago?"

Few people can answer this question exactly.

In contrast, "What was the name of your first love?"

Few people cannot answer this question. Almost everybody can remember who their first love was, even if this memory is from primary school or junior high school. Nobody can forget this! You can appreciate how such things are memorized. The first love is not remembered by name. Instead, her voice, smiling face, skin, or the smell of her hair in the classroom or the school playground are remembered. You can remember the emotions that you felt when you talked with her. This memory is the combination of all fragment memories, saved as the connection of many parts. This is a typical long-term memory.

By the way, did you ever cram for a test the night before in your school days? Although this method of study appears efficient, it is not efficient at all. For example, when you see world history, you cannot remember the names of Greece's three great tragedy poets. Neurologically speaking, their names were in your short-term memory, but since they were fragmental (failed to transform into long-term memory), you had forgotten this information by the next day.

If you have these fragmental memories, connected with other information such as a historical map of the Greek world or portrait of the poets, in your memory, such as the memories of the first love, you can easily remember the names of three Greek poets. In addition, the relationship between the three persons and how they were working can be memorized in connection with the story of the poet and places on the historical map. By this method, you can remember the three names of the poets. This is the power of long-term memory. This power cannot be obtained by cramming the night before a test.

Let's consider the "King of the Quiz," who can answer every question. He can answer the whole story from fragmental information. This ability is a little different from the intelligence. He has an effective methodology to convert many and fragmental short-term memories to his long-term memory. He naturally developed the methodology by himself, or his parents or a teacher taught him the method and he earnestly equipped the method. He has the names of three Greek poets in memory in connection with other types of information, such as the memories of your first love. Therefore, if you learned this method on your own, you may defeat the King of Quiz. First, you should investigate the information on the three poets.

We must consider how short-term memory and long-term memory differ from the perspective of neuroscience. This is due to the difference in the areas of the brain that save these memories. Long-term memories are saved in the neocortex, the largest portion of the brain. The neocortex comprises six overlapping layers; it has accumulating cylindrical structures connected named as a "neural network." Each cylinder corresponds to a fragmental memory (a simple concept such as a specific triangle or smell) and the information is connected by neural networks to form long-term memory.

6.1.2   Memory on a Single Layer

Well, how does long-term memory form in the neocortex? Even now this is one of the mysteries of neuroscience. One possible answer is stringent emotion. For example, everybody feels strong emotions when speaking with a first love. Thus, you remember all the surrounding things, not only her face and body, but also her voice, smell, and hands by the power of long-term memory. These long-term memories are kept by the vast neural connection between cylindrical structures of the neocortex. Since long-term memories are formed by vast connections, even when some neurons are partly damaged, the brain itself can keep the memory. Even patients with severe dementia may have long-term memories.

In contrast, the hippocampus, responsible for short-term memories, is located below the neocortex composed of six layers and has the shape of a thin, single, rolled-up paper (Fig. 6.1). Please imagine large neurons lining up in an ordered manner on cylinder-like thin paper. Since the hippocampus is like a thin, single paper, it cannot form column-like structures. Fragmental information can be connected, but the density is quite different from the neocortical connections. Short-term memory has structural limitations and is easily lost. Within a short time, this memory is replaced by a new one. This is a neurological explanation of forgetfulness. In addition, this is the physiological reason why we cannot achieve perfect answers by cramming the night before a test [1, 2].

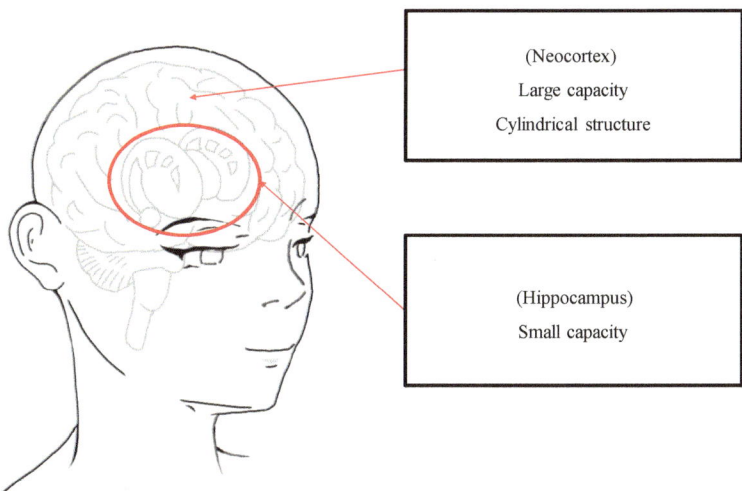

Fig. 6.1 Long-term memory (neocortex) and short-term memory (hippocampus). The brain stocks long-term memories in the neocortex and short-term memories in the hippocampus. Note that the neocortex saves a large amount of information in the vast number of cylinder-like structures, in contrast, the hippocampus can save a very small amount of information on a thin layer-like structure

6.2   What Happens in the Alzheimer's Brain?

6.2.1   Initiation of Dementia by Energy Deficiency of Hippocampal Pyramidal Neurons

If you ask a person "What did you eat for breakfast?" and he cannot answer this question, it may be possible that he has dementia. It is just a possibility. If he forgets that he ate breakfast, but he did, he likely has dementia. Dementia is no longer a rare illness. Indeed, 25% of readers of this book will develop dementia. In addition, the lives of the same population will be affected by dementia in their close friends or family. Overall, half the readers of this book will be seriously affected by dementia (either themselves or close friends or family). Thus, we have to consider what action we should take to prevent dementia in the future. Even a significant delay in the progression of dementia may greatly improve quality of life. However, many people think that they have to give up the future. If Alzheimer's disease (AD) has begun to progress, what can be done is to cause a small delay in some drugs. Is this right? The author will consider a possible countermeasure to combat AD in terms of brain energy. The biggest advantage of this countermeasure is that anybody can adopt this in his daily life.

To fix specific information as short-term memory in pyramidal neurons of the hippocampus, the neurons need complex mechanisms and demand a huge volume of energy. Pyramidal neurons responsible for short-term memory have N-Methyl-D-Aspartate (NMDA) receptors, a cell surface receptor for glutamate (a neurotransmitter). As glutamate is an excitatory neurotransmitter, its binding to the NMDA receptor induces a large influx of Ca^{2+} ions into the synapses on dendrites. Finally, this event leads to the transcription of a large number of proteins. This is why pyramidal neurons are remarkably large and demand a huge volume of energy compared with other granular neurons. In this background, when sufficient energy is not supplied to the brain due to aging, pyramidal neurons are the most seriously impacted by energy deficiency.

6.2.2   Dementia and Type 3 Diabetes

Many researchers suppose that the first event of dementia is energy deficiency in the pyramidal neurons of the hippocampus. But why are hippocampal pyramidal neurons intensively deficient in energy? Since many valuable studies have tried to answer this question, I will provide an ordered explanation. Many researchers suggest that dementia is diabetes in the brain. In particular, it is reported that patients with type 2 diabetes may have dementia. Therefore, people with high blood glucose levels are advised to read this section carefully. However, please don't worry too much, as I will present to you a possible countermeasure.

First, let's look into the actions of insulin in the brain (Fig. 6.2). Insulin is a hormone that reduces blood glucose. Most cells have GLUT4, a glucose transporter activated by insulin. Thus, the activation of GLUT4 is key to the reduction of glucose by insulin. GLUT4 has the unique function of permitting glucose intake only when there is insulin. When insulin reaches cells and binds to the insulin receptor, GLUT4 opens and increases the intake of glucose from blood to cells and, finally, glucose is rapidly transferred to skeletal muscle or fat tissue. This is why an increase in glucose leads to a rapid reduction in blood glucose [3].

Insulin has a clear communication (signal transduction) between insulin and GLUT4 to reduce blood glucose levels, as shown in Fig. 6.2. Thus, insulin does not effectively reduce blood glucose when this communication is not working well. Type 2 diabetes is a disease due to a disorder of this communication ("Signal transduction" in Fig. 6.2). Actually, somatic cells other than pyramidal neurons in the brain do not have serious outcomes, even when glucose intake is slowed [4, 5].

However, neurons, especially pyramidal neurons in the hippocampus, have a completely different story. After all, pyramidal neurons demand extraordinarily large amounts of energy. Disorder of glucose intake in pyramidal neurons will lead to serious problems. This happens in the brain with mild cognitive impairment, such as in early dementia. "Type 3 diabetes" has recently been proposed, because the same process as type 2 diabetes has been shown to occur in the brain of patients with dementia [5].

Why are pyramidal neurons in the hippocampus sensitive to energy deficiency? Other pyramidal neurons also have a large energy demand. Pyramidal neurons of the hippocampus and neocortex demand extraordinarily large energy, but pyramidal neurons in the hippocampus have a unique and specific story. Pyramidal neurons of the hippocampus have a different glucose uptake gatekeeper from those of other pyramidal neurons. Figure 6.3 explains the distribution of GLUT1 and GLUT4 in the hippocampus. GLUT1 is expressed in astrocytes and GLUT4 is expressed in pyramidal neurons [4, 5].

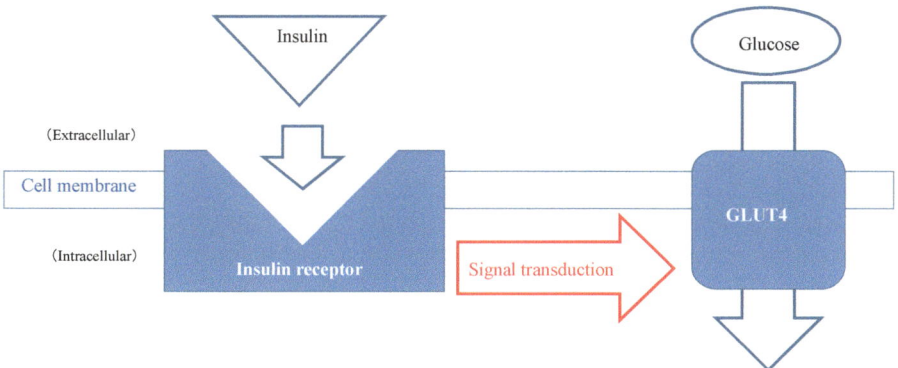

Fig. 6.2 Basic function of insulin [3]. Insulin enhances glucose uptake through GLUT4 utilization. Signal transduction is essential for glucose uptake and its breakdown leads to type 2 diabetes. Dementia is defined as "Type 3 diabetes," where signal transduction becomes weak. GLUT4 cannot fully work on hippocampal pyramidal neurons. Therefore, dementia initiates an energy shortage of the pyramidal neurons. This is why ketone bodies can rapidly rescue pyramidal neurons by supplying energy to neuronal mitochondria

Astrocytes   Pyramidal neurons

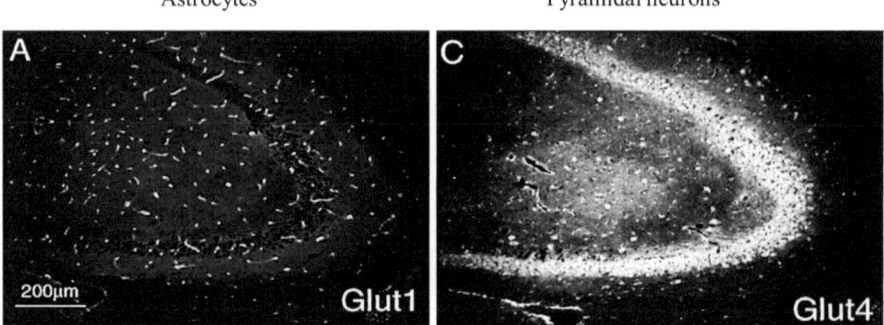

Fig. 6.3 Expression of GLUT4 on hippocampal pyramidal neurons [6]. This suggests that type 3 diabetes may lead to an energy shortage of hippocampal pyramidal neurons. Pyramidal neurons of the hippocampus are highly distinctive from astrocytes and other pyramidal neurons such as in the neocortex. This is highly related to the early onset of dementia due to energy shortage when glucose uptake through GLUT4 is down in diabetes

6.2.3 Hippocampal Pyramidal Neurons with a High Dependency on Insulin

Pyramidal neurons of the hippocampus have GLUT4 and astrocytes in the hippocampus have GLUT1. GLUT4 is active only when insulin is present, whereas GLUT1 is always active. Therefore, astrocytes can maximize glucose uptake (Fig. 6.3). In contrast, pyramidal neurons can take up glucose only when there is insulin. Pyramidal neurons of the hippocampus will have serious energy problems when insulin is not performing well (for example, in type 2 diabetes). They will fall into energy deficiency because of impaired glucose uptake. This is partly because of their extraordinary energy demands and partly because they express GLUT4.

Pyramidal neurons are uniquely arranged in the hippocampus. In comparison with pyramidal neurons in the neocortex, they are densely packed in a very narrow area, as shown in Fig. 6.4. It appears that the hippocampus has invented ways to increase its functionality to perform highly important and complex tasks (short-term memory), even though it has limited capacity compared with the neocortex. However, this is just next to danger, because of the dense, highly stress-sensitive pyramidal neurons, although this invention increases its functionality to the maximum level [7].

Please imagine what their destiny is under stress that exceeds their capacity. This may lead to a situation in which all of the dense pyramidal neurons will die. The most important thing for planning complex tasks, such as in the brain, is to prepare for the worst situation. However, the human brain does not prepare for the worst situation. This is unbelievable but true. Alternatively, this may be a strategy to protect the neocortex (long-term memory) at the expense of the hippocampus (short-term memory). It is impossible to know what the strategy is. We must always conduct prevention and vigilance in order not to cause such a situation. We must plan an effective countermeasure by knowing the weak point of the brain [7].

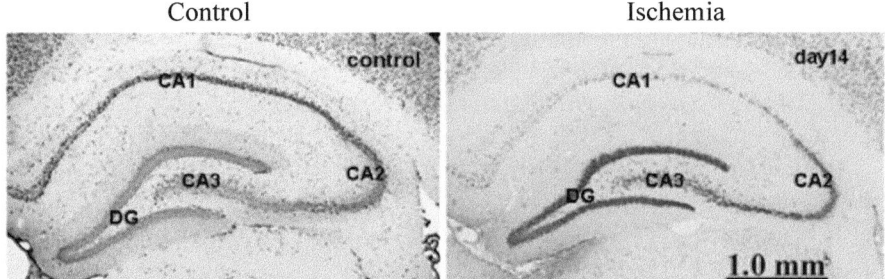

Fig. 6.4 Pyramidal cell death of hippocampus [7]. Pyramidal neurons in the hippocampus are all gone but other cells are not damaged. Note that these cells are highly sensitive to stress, which may lead to the onset of dementia. This volatility of the pyramidal neurons of the hippocampus must be closely related to the onset of dementia

The right image in Fig. 6.4 shows the worst situation. The authors show the special impact of global ischemia, an experimental model of brain stroke, on the mouse brain. CA1 is a special arrangement of pyramidal neurons responsible for short-term memory, suggesting that they are more vulnerable to stress than other pyramidal neurons. It is notable that only the pyramidal neurons of the hippocampus are completely gone, whereas the other cells are intact. In addition, global ischemia puts the whole brain under stringent stress but the damage concentrates specifically on the pyramidal neurons of the hippocampus (Fig. 6.4). As the hippocampus contains highly densely arranged stress-sensitive pyramidal neurons, short-term memory is gone within a short time (several days). This damage is irreversible and thus, he has to live a life without short-term memory.

We need just a little care and effort to protect the pyramidal neurons of the hippocampus, as you can see from above. We should be conscious of the situation that our brain is always under threat. The brain must live with its weak points. We have to express special thanks to the brain every day before going to bed because it made it through the day safely.

Therefore, let us take a second look and consider a possible countermeasure. First, to summarize, the hippocampal pyramidal neurons have the following features [1, 2]:

1. They are arranged in lines on a cylinder-like structure.
2. They depend on insulin-sensitive glucose uptake.
3. Their energy deficiency is linked to the loss of short-term memory.

6.2.4   Recovery of Neuronal Functions from Crisis

Next, we should look into a possible method to avoid neuronal crisis. Hippocampal pyramidal neurons will be subjected to the following situations in early dementia, even if they do not recognize the symptoms. The first step to an effective countermeasure is to determine the real condition surrounding the hippocampus.

1. Neurons are intact without any death or damage.
2. Neurons are ready to process short-term memory.
3. Neurons experience some trouble in the communication between insulin and GLUT4 (insulin resistance) and suffer from energy deficiency due to the decrease in glucose uptake.
4. The addition of glucose (carbohydrate intake) is nonsensical to pyramidal neurons because glucose uptake depends on insulin.
5. The addition of ketone bodies may recover the functions of pyramidal neurons.

Therefore, it is now the turn of the ketone bodies, another energy substrate.

As you will have noticed, the purpose of this book is to inform you of another energy substrate. One of the aims of writing this book is to let you know that brain aging is slowed notably by the use of ketone bodies, another energy substrate. To maintain brain health and longevity, we have to take a little more care of ketone bodies. The ketone bodies are energy substrates that can replace glucose, like a special rescue, to save the brain when the brain is out of order due to energy deficiency. Caring for ketone bodies is a great help to the health and longevity of the brain and the whole body. This is the aim of writing this book. Further, ketone bodies may perform highly effective rescue of the brain when it is not working well. Living a daily life without knowing this is too good an opportunity to waste, isn't it? I wish to ask you this question.

6.3 Science of Alzheimer's Disease

6.3.1 Effective Countermeasures to Protect Neurons Affected by Energy Shortage

The hippocampus is mainly responsible for short-term memory and has the following features compared with the neocortex.

1. Low capacity
2. Dense arrangement of each pyramidal neuron
3. Expression of GLUT4

I have explained these points in a brief review:

1. Because of its small capacity, it cannot save large amounts of information. The newly saved information (short-term memory) is gone within several minutes or several hours unless it moves to the neocortex and is saved as long-term memory.
2. Although the dense arrangement of pyramidal neurons can increase their functionality, it poses a serious risk that all pyramidal neurons may be lost at the same time.
3. Pyramidal neurons of the neocortex expressing GLUT3 do not need the action of insulin when they take up glucose. In contrast, pyramidal neurons in the hippo-

campus expressing GLUT4 need insulin. They cannot absorb glucose in the absence of insulin.

Because of this background, even a healthy old person has a sparse arrangement of pyramidal neurons (CA1) in the hippocampus. It may happen that, with age, we often forget things and hesitate to take them into memory.

6.3.2   Effective Countermeasures Against Alzheimer's Disease

In Alzheimer's disease, which accounts for 70% of all dementia patients, the amyloid hypothesis is well established. This hypothesis states that beta-amyloid, a protein, is involved closely with neurons and induces neuron degeneration. Since the precipitation of beta-amyloid and energy deficiency can mutually enhance each other, the brain may fall into a negative spiral when neurons are exposed to both stimuli (Fig. 6.5) [8].

Of course, it is reasonable that other factors are driving the brain into dementia than energy deficiency. Even in the case that another factor induces dementia, eliminating energy shortage by supplying another energy substrate is important to break the negative spiral into dementia as shown in Fig. 6.5. Thus, the supply of another energy substrate is important in the early stages of chronic neuronal degeneration [8].

Is the method of eliminating the energy shortage by supplying ketone bodies, which the author is proposing here, effective in Alzheimer's disease? As reported by many papers and books, it appears to be effective. For example, Mary T. Newport reported in "Alzheimer's Disease: What If There Was a Cure?" that a small increase in ketone body concentration (0.2–0.5 mM) was highly effective in Alzheimer's

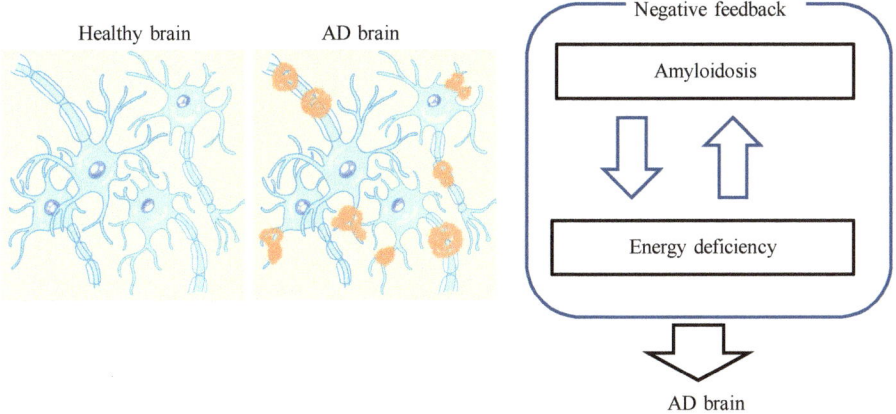

Fig. 6.5 Negative feedback loop of the AD brain [8]. Accumulation of beta-amyloid enhances energy deficiency and energy deficiency induces the accumulation

disease [9]. This suggests that small ketogenic (0.2–0.5 mM) is highly effective although stringent carbohydrate restriction is not necessary.

The range of ketone body concentrations is designated as "small ketogenic" as an effective countermeasure against Alzheimer's disease. Those concentrations can significantly activate the HCAR2 receptor as described in Chap. 7. In addition, old people in the villages of longevity may maintain these ketone body concentrations estimated from their eating habits.

When one is anxious as to whether they are in the early phase of dementia, this information may provide a ray of hope. The anxiety may end with "it was my fault" and is dealt with calmly. Ketone bodies have attracted my attention since I read Mary T. Newport's book. I read it over and over again until the book became tattered. Since ketone bodies have emerged as one of the major research subjects, I am grateful to Mary T. Newport [9].

Pyramidal neurons of the hippocampus may fall into energy shortage in the early stages of dementia. By supplying an abundance of energy to the brain, it can find a way to save itself or delay the progression of dementia. At least, I believe this story is right. Therefore, we have to depend on the power of ketone bodies as well as glucose.

Why do ketone bodies preserve cognitive functions by rescuing pyramidal neurons from energy shortage? As pyramidal neurons of the hippocampus have a central role in short-term memory, ketone bodies can enhance memory by restoring the energy supply. Ketone bodies act directly on the mitochondria of pyramidal neurons. Pyramidal neurons in the hippocampus have numerous spines on their dendrites and the spines are synapses receiving input from the surrounding granular neurons. As neurotransmission in the synapses requires an energy supply, mitochondria accumulate in both presynaptic and postsynaptic cells.

The cognitive functions of mice are significantly recovered by the oral administration of ketone bodies. Ketone bodies are absorbed in the digestive ducts, targeted to the microvessel of the brain, and pass through the blood-brain barrier. Astrocytes take up the ketone bodies and pass it through the monocarboxylic acid transporter (MCT) to pyramidal neurons in the hippocampus. Ketone bodies can translocate into the mitochondria and are metabolized to produce the energy substrate ATP. Through these pathways, ketone bodies can act directly on neuronal mitochondria. The pyramidal neurons of the hippocampus can enhance cognitive function by using ketone bodies, even when neurons are functioning under low glucose.

6.4 Synaptic Transmission in Detail

Since the synaptic transmission process needs a lot of energy, mitochondria are highly accumulated to provide abundant energy. Synaptic transmission demands large energy and for this demand, many mitochondria are accumulated in postsynaptic neurons as shown in Fig. 6.6. An electron microscopy image of the synapses shows that mitochondria exist in the synapses. Astrocytes contain mitochondria but

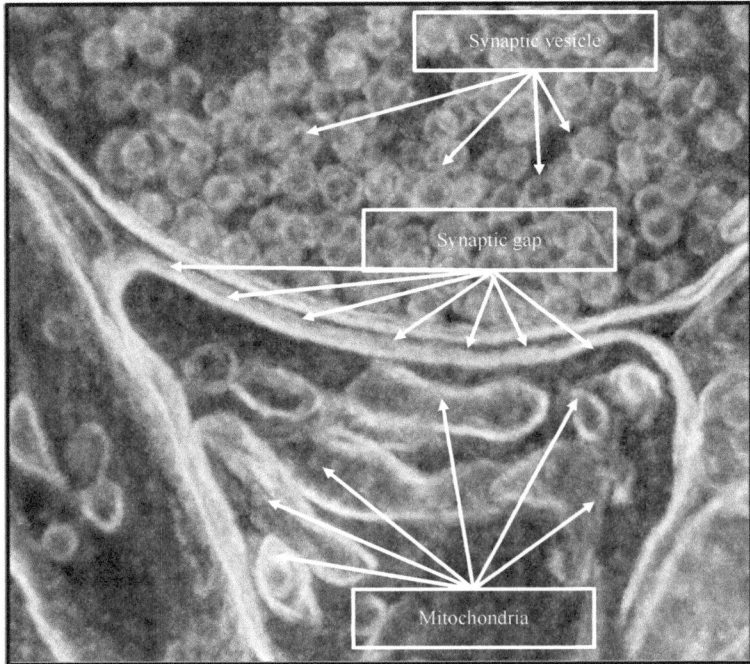

Fig. 6.6 Accumulation of mitochondria in synapsis. Several mitochondria are accumulating in the postsynapses because of huge energy requirements, suggesting that huge energy is demanded for synaptic transmission. Therefore, the function is greatly affected by energy shortage. It is believed that one of the first visible changes during the progression of dementia is a decrease in synapses

at a much lower density than neurons. Therefore, ketone bodies can directly act on these mitochondria and preserve neuronal functions when blood glucose levels are fluctuating. This is why cognitive function is recovered as soon as ketone bodies are orally administered.

Although the brain usually uses glucose for the production of the energy substrate ATP, the presence of insulin is critically required when pyramidal neurons take up glucose, as mentioned. This is a highly special point of hippocampal pyramidal neurons, in contrast, cortical pyramidal neurons do not require the presence of insulin. This is why a decline in short-term memory is highly associated with type 2 diabetes. In contrast, ketone bodies do not need the presence of insulin and target the mitochondria. Thus, it has high energy efficiency. In addition, ketone bodies just pass through astrocytes, although glucose has to be metabolized to lactate.

The essence of neurotransmission is a transmission of neuronal information mediated by small compounds such as glutamate but not by electrical action potentials. The brain uses a substance named "glutamate" as one of the major excitatory neurotransmitters and plays a central role in short-term memory formation. Here, I explain the neurotransmission of glutamate and provide basic information on what a synapse is in Fig. 6.7. The synapse has a gap between presynaptic nerves and

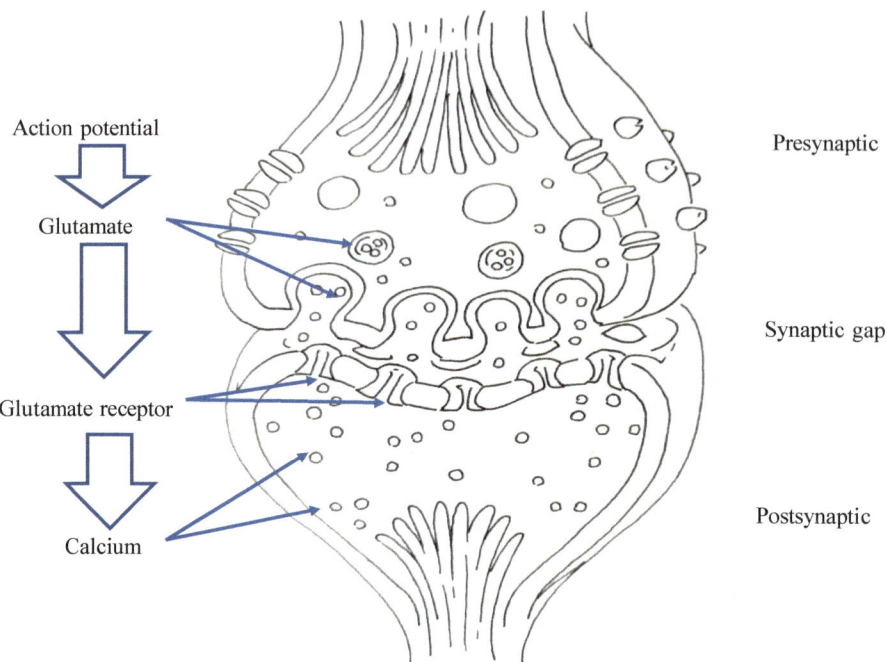

Fig. 6.7 The essence of synaptic transmission is energy conversion, action potential (electrical energy)-glutamate (chemical energy)-calcium ion (electrical energy) [10]. Since the postsynaptic event (glutamate receptor to calcium) needs much energy, mitochondria are accumulated in postsynaptic neurons. The structure of a dendrite of excitatory postsynaptic neurons is called as "spine" as described in Fig. 6.9

postsynaptic nerves. Neither of the nerves attach and there is a clear gap. Presynaptic and postsynaptic nerves correspond to the neuronal axon terminal and dendritic spine, respectively.

Here, the details of neurotransmission are presented to provide a brief understanding of synaptic transmission between a presynaptic neuron and a postsynaptic neuron. The basic function of synapses is the informational conversion of action potentials of presynaptic nerve terminals into cellular function (increase in the intracellular Ca^{2+} ion concentrations) of postsynaptic nerves. Conduction is the transmission of electrical signals mainly mediated by Na^+ current. However, electrical signals cannot directly transmit to postsynaptic nerves because the synapsis is completely insulated by a small gap between the presynaptic and postsynaptic nerves. Electrical signals cannot pass through this gap. Energy conversion to chemical information is required for transmission into postsynaptic nerves. Neurotransmitters are a means of chemical energy for transmission across the gap. For example, glutamate is used as a neurotransmitter.

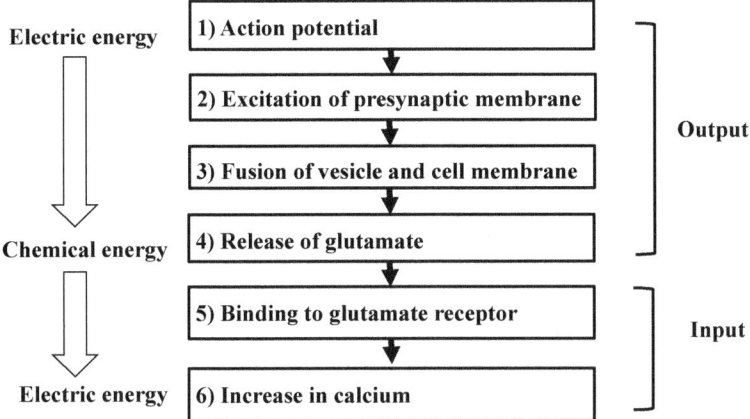

Fig. 6.8 Sequenced events of synaptic transmission [10]. Note that the essence of synaptic transmission is energy conversion. Here, sequential energy conversion occurs by electric energy: action potential (1~3)→chemical energy: glutamate (3~5)→electric energy: calcium current (6)

Energy conversion by use of high-level techniques occurs in the synapsis as shown in Fig. 6.8. The flow of energy status can be demonstrated as action potentials (electric energy) → glutamate (chemical energy) → Increase in Ca^{2+} (electric energy). Mitochondria accumulate around the synapsis in postsynaptic neurons to supply the energy substrate ATP to maintain the high performance of synaptic transmission. (Astrocytes wrap around the synapsis to support synaptic transmission.) An abundant supply of energy (lactate and ketone bodies) from astrocytes enables neurons to maintain high-performance synaptic transmission. Although the brain is one of the most energy-demanding organs, neuronal synapsis has the highest energy demand among the parts of the brain.

Let us continue to explain the system of synaptic transmission. Action potentials are initiated from the soma and reach the synaptic terminal of the presynaptic nerve through axons. By excitation of the presynaptic nerve membrane, intracellular Ca^{2+} increases, and vesicles, including glutamate, are released into the synaptic gap by fusing with the cell membrane. The glutamate released into the gap binds to a specific glutamate receptor (NMDA receptor) on the postsynaptic nerve. By this stimulation, intracellular Ca^{2+} increases, and short-term memory is finally formed by the Ca^{2+}-mediated signals.

Briefly speaking, the synapsis provides a space for the system that accepts the action potentials of neurons, conducts membrane fusion, releases neurotransmitters and receptor binding, and transfers information into a postsynaptic nerve. Thanks to this system, neuronal information can be transferred between neurons without clogging and stopping. Electrical signals that have some information can move to the target organ thanks to this complex system of synaptic transmission.

6.5   The NMDA Receptor on Synapsis

6.5.1   Short-Term Memory in the Hippocampus

Many scientific reports indicate that activation by glutamate receptors is the first event leading to the formation of short-term memory in pyramidal neurons of the hippocampus. Here, we have to see the neurotransmission mechanism of the glutamate synapse of hippocampal neurons. First, it is necessary to explain the mechanism through which short-term memory is established in the synapses of CA1 pyramidal neurons of the hippocampus. This mechanism requires NMDA receptor (a type of glutamate receptor) activation on the spines of neuronal dendrites (Fig. 6.9) [11–13].

Glutamate released from presynaptic cells binds to the NMDA receptor on spines of dendrites and transmits information to postsynaptic cells, as shown in Fig. 6.9. By activating such a mechanism, the NMDA receptor becomes activated and allows Ca^{2+} to flow into postsynaptic cells. This increase in Ca^{2+} is essential to establish short-term memory [11–13].

Excitation of presynaptic cell membrane → Glutamate release → NMDA receptor activation → Ca^{2+} influx of postsynaptic cells → Short-term memory

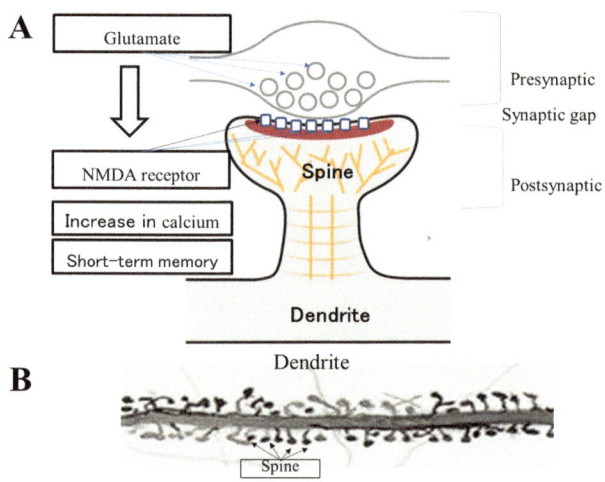

Fig. 6.9 Glutamate induces short-term memory through NMDA receptor (a type of glutamate receptor) activation (**a**). Spines on neuronal dendrites of pyramidal neurons are the NMDA-activated synapse. Note that glutamate induces short-term memories, but in emergencies, it may initiate neuronal damage by overloading of Ca^{2+} ions. It is well-known that the structures of spines are always changing, and spines are easily lost when they are not used for a long time. Spines are particle-like structures on dendrites (**b**). Thousands of spines are on dendrites to receive information from other neurons and have NMDA receptors on the postsynaptic membrane in pyramidal neurons of the hippocampus

There is a big problem with this sequence of events. For unknown reasons (stringent stress, etc.), excessive glutamate release is problematic because it leads to an excessive influx of calcium ions.

6.5.2   Possible Method to Protect the Brain from Neuronal Death

When neurons are overloaded with Ca^{2+} by activation of NMDA receptor, they cannot live anymore. Consequently, short-term memory is no longer valid. In technical terms, "excitotoxicity" has been induced. In neurons, excessively high concentrations of Ca^{2+} are highly toxic or cause death. In the Alzheimer's brain, excessive glutamate release is reported to occur. This is a vicious circle. However, potential help for the brain is hidden in this system. In other words, there is suggested to be a potent inhibitor of this stream that can stop the sequential events leading to AD and save the brain. The inhibitor can suppress neuronal cell death by global ischemia, a highly stringent stress, in an experimental mouse model. For example, the inhibitor can fully suppress the death of pyramidal neurons in the hippocampus, as shown in Fig. 6.10. In this case, there is a sequence of events leading to the death of pyramidal neurons.

Global ischemia → excessive glutamate release → excessive NMDA receptor activation → excessive Ca^{2+} ion influx → death of pyramidal neurons [7]

May I ask you which event is inhibited by the inhibitor? It is in this process.

Inhibition of NMDA receptor activation may inhibit the death of pyramidal neurons.

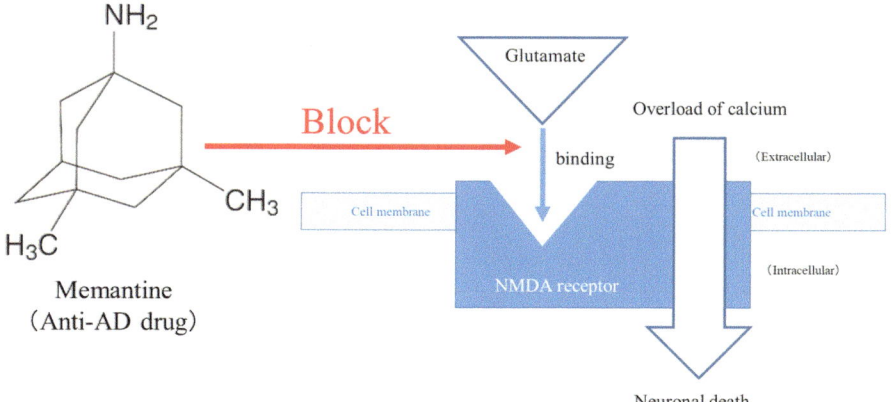

Fig. 6.10 Memantine and NMDA receptor [11–13]. Glutamate may induce neuronal death by Ca^{2+} overload. Therefore, memantine, an NMDA antagonist, can function as an anti-AD drug. Memantine is just a case of a successful NMDA antagonist among vast numbers of candidates developed by major pharmaceutical companies

Many pharmaceutical companies and universities have been competing for an effective inhibitor of NMDA receptor activation, based on the hypothesis that it can inhibit AD progression (Fig. 6.10). Finally, this hypothesis has been proved to be right.

The group of Stuart A. Lipton (The Scrips Research Institute, La Jolla, CA) identified memantine as an NMDA antagonist with minimal side effects, and it has since been used for the treatment of patients with AD. As there are several anti-AD drugs other than memantine, it is clear that this research is valuable [11–13]. In addition, this drug is already commercialized in many countries for clinical use.

6.6 Energy Deficiency in Pyramidal Neurons

6.6.1 Changes in Energy Substrates During the Progression of Alzheimer's Diseases

Figure 6.11 presents a schematic representation of the route leading to dementia. In many cases, a starting point is a constitutive energy deficiency in pyramidal neurons [8]. During *Stage I*, the structural degeneration of neurons such as a decrease in

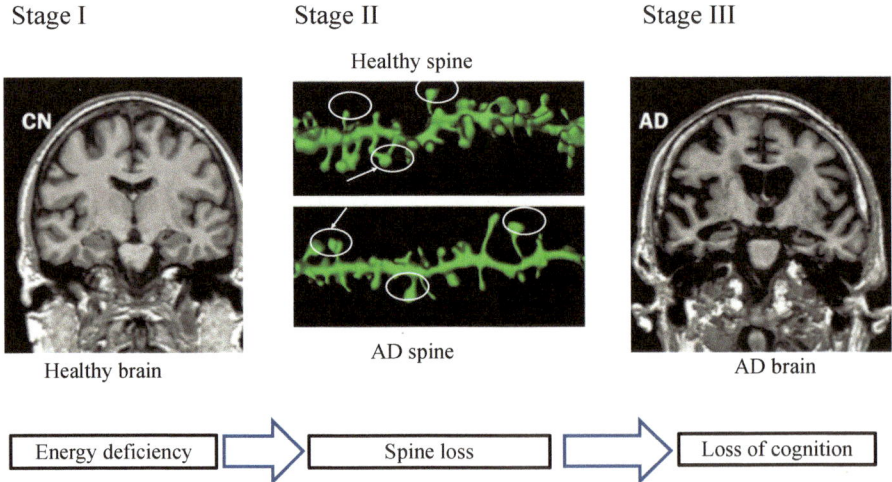

Fig. 6.11 Changes of pyramidal neurons in the AD brain [14, 15]. In *Stage I*, neurons do not degenerate yet, but pyramidal neurons are continuously suffering from energy shortage. During this stage, ketone bodies can rescue neurons during the phase of energy deficiency. In *Stage II*, neurons start to degenerate. Spines are firstly lost, and neuronal functions are inhibited. Still in this phase, ketone bodies can partly recover brain functions such as cognitive functions. In *Stage III*, neurons have degenerated to die, and a large space is observed in the hippocampus as well as in the neocortex. Note that *Stages I, II*, and *III* in this figure are those in Fig. 6.12

spines and dendrites has not yet begun. Thus, nothing may happen if there is another energy supply (ketone bodies). However, during *Stage II*, if this deficiency is left unattended, dendrites and spines begin to decrease, and neuronal functions begin to break down. The degeneration progresses further and, finally, the neocortex shrinks significantly during *Stage III*. The most effective countermeasure is to supply ketone bodies, an alternative energy substrate, in the first step in which neurons are demanding energy but not yet damaged.

Figure 6.12 shows a schematic of the contribution of each energy substrate in parallel to the progression of dementia. For the simplest explanation, it can be divided into three stages.

In *Stage I*, insulin secretion increases to compensate for the decrease in insulin activity. In response, glucose uptake increases. By increasing insulin secretion, insulin effects are strengthened. In technical terms, this is known as "insulin resistance." *Stage I* is the insulin resistance stage.

Stage I is the time at which insulin resistance begins. Diabetes reserve is in *Stage I* or about to be in *Stage I*. Insulin-secreting cells receive too great a load, which cannot continue for a long time.

Stage II is the time when the system degenerates day by day. Real diabetes begins. Clinical symptoms may occur in a chain reaction in the brain. Pyramidal neurons do not work well under conditions of energy deficiency; although astrocytes and hepatic cells try to compensate for the deficiency, the functions of neurons are declining.

In *Stage III*, since the uptake of glucose is inhibited, functions of pyramidal neurons (short-term memory) are declined. However, in place of glucose, the supply of ketone bodies does not support their functions. It is therefore highly possible that the brain function is irreversibly degenerated. Countermeasures should have been taken in *Stage I* or *Stage II*.

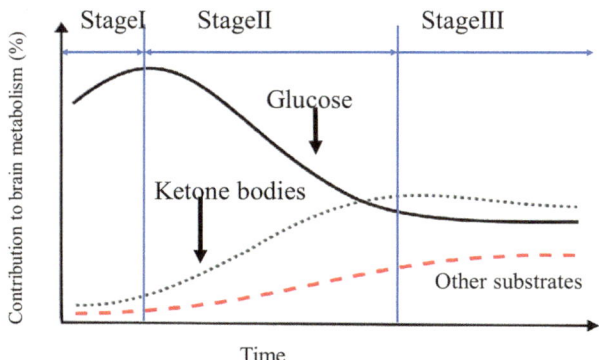

Fig. 6.12 Changes of energy substrates during the AD progression [16]. Note that ketone bodies are not yet fully increased. Thus, AD does not progress if ketone bodies are activated in *Stage I*. The important is a practical method to activate ketone bodies in *Stage I*. The small ketogenic is proposed here to fill the demand and this is the purpose of this book

6.6.2 Ketone Bodies to Combat the Decline in Synaptic Functions

In *Stage I* and *Stage II*, ketone bodies are an effective countermeasure against dementia. This is because ketone bodies are a much better energy substrate in terms of mitochondrial activation. Low concentrations of ketone bodies (0.2–0.5 mM) may prevent or suppress dementia as described by Mary T. Newport in her book [11]. The countermeasure is more effective when it is taken earlier. In this case, ketone bodies have been so effective in many clinical tests that patients with Alzheimer's disease are quickly recovering their cognitive functions. In *Stage I* (Fig. 6.12), the countermeasure (*small ketogenic*) is highly effective. The potential patients may not enter Stage II and Stage III if they curb excessive carbohydrate intake and increase ketone bodies to a very small extent.

As shown in Fig. 6.13, the spines are numerous thorn-like structures on the dendrites of neurons, and these are the apparatus to receive input from other surrounding neurons. Each spine is a synapse. Healthy, young persons have many spines on each neuron. However, the number of spines gradually declines with age. The decline of spines is well correlated with impairment of cognitive functions as shown in the Stage II of Fig. 6.12.

As mentioned above, the total amount of information declines with a decreasing number of spines. Particularly, in the progression of dementia, the number of spines decreases. The damage to the whole brain leads to a lower input from surrounding granular neurons. As mentioned above, a pyramidal neuron is like an informational hub. As the input to the spines declines, the output from the axons decreases. (As an analogy, it is like a quiet and lonely entrance of the hub Narita or Haneda airport as

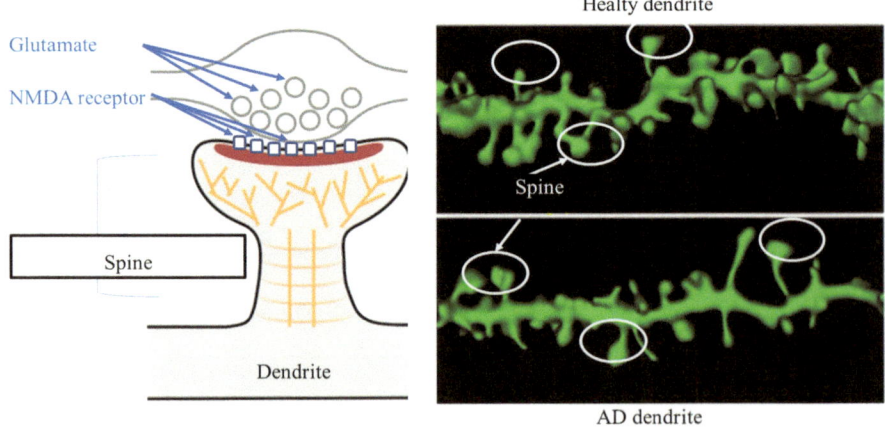

Fig. 6.13 Spine degeneration in the brain of Alzheimer's disease [15]. Spines on dendrites of hippocampal pyramidal neurons are lost in the AD brain, which may lead to impairment of short-term memories. AD progression initiates an energy shortage if left unattached for a long time. The first event in the progression is the degeneration of spines (synapsis) on hippocampal pyramidal neurons

usage of domestic terminals continues to decrease. However, there may have been many countermeasures that could be implemented to not lead to such a situation.)

References

1. English DF, McKenzie S, Evans T, Kim K, Yoon E, Buzsáki G. Pyramidal cell-interneuron circuit architecture and dynamics in hippocampal networks. Neuron. 2017;96(2):505–520.e7.
2. Soltesz I, Losonczy A. CA1 pyramidal cell diversity enabling parallel information processing in the hippocampus. Nat Neurosci. 2018;21(4):484–93.
3. Jaldin-Fincati JR, Pavarotti M, Frendo-Cumbo S, Bilan PJ, Klip A. Update on GLUT4 vesicle traffic: a cornerstone of insulin action. Trends Endocrinol Metab. 2017;28(8):597–611.
4. Szablewski L. Glucose transporters in brain: in health and in Alzheimer's disease. J Alzheimers Dis. 2017;55(4):1307–20.
5. López-Gambero AJ, Martínez F, Salazar K, Cifuentes M, Nualart F. Brain glucose-sensing mechanism and energy homeostasis. Mol Neurobiol. 2019;56(2):769–96.
6. Leto D, Saltiel AR. Regulation of glucose transport by insulin: traffic control of GLUT4. Nat Rev Mol Cell Biol. 2012;13(6):383–96.
7. Yasuda Y, Shimoda T, Uno K, Tateishi N, Furuya S, Tsuchihashi Y, Kawai Y, Naruse S, Fujita S. Temporal and sequential changes of glial cells and cytokine expression during neuronal degeneration after transient global ischemia in rats. J Neuroinflamm. 2011;8:70.
8. Camandola S, Mattson MP. Brain metabolism in health, aging, and neurodegeneration. EMBO J. 2017;36(11):1474–92.
9. Newport MT. Alzheimer's disease: what if there was a cure?: The Story of Ketones. Basic Health Publications; 2011.
10. Barnes JR, Mukherjee B, Rogers BC, Nafar F, Gosse M, Parsons MP. The relationship between glutamate dynamics and activity-dependent synaptic plasticity. J Neurosci. 2020;40(14):2793–807.
11. Xia P, Chen HS, Zhang D, Lipton SA. Memantine preferentially blocks extrasynaptic over synaptic NMDA receptor currents in hippocampal autapses. J Neurosci. 2010;30(33):11246–50.
12. Lipton SA. Pathologically activated therapeutics for neuroprotection. Nat Rev Neurosci. 2007;8(10):803–8.
13. Zhao Y, Navia BA, Marra CM, Singer EJ, Chang L, Berger J, Ellis RJ, Kolson DL, Simpson D, Miller EN, Lipton SA, Evans SR, Schifitto G, Adult Aids Clinical Trial Group (ACTG) 301 Team. Memantine for AIDS dementia complex: open-label report of ACTG 301. HIV Clin Trials. 2010;11(1):59–67.
14. Vemuri P, Jack CR Jr. Role of structural MRI in Alzheimer's disease. Alzheimers Res Ther. 2010;2(4):23.
15. Chang PK, Boridy S, McKinney RA, Maysinger D. Letrozole potentiates mitochondrial and dendritic spine impairments induced by β amyloid. J Aging Res. 2013;2013:538.
16. Neth BJ, Craft S. Insulin resistance and Alzheimer's disease: bioenergetic linkages. Front Aging Neurosci. 2017;9:345.

Chapter 7
Small Ketogenic Rescues the Brain

Abstract As Mary T. Newport reported in the book, *"Alzheimer's Disease: What If There Was a Cure?"* that cognitive functions of dementia will significantly recover by increasing ketone bodies to a small extent. Her husband had early-onset dementia [Newport (Alzheimer's disease: what if there was a cure?: The story of ketones. Basic Health Publications, 2011)]. She had searched patient information by herself and found that ketone bodies were a little increased and that even these levels of ketone bodies were effective against the progression of dementia. Then, she added coconut oil 30 g/day to his meals. Just after her husband began to have this, she found that his cognitive functions were greatly recovered. It has been a wonderful gift to scientists that she had measured the concentration of ketone bodies of her husband although back then it was not easy to measure ketone bodies in the blood. She reported that the concentrations of ketone bodies were increased from 0.1 mM to 0.2–0.5 mM. Surprisingly, these levels of ketone bodies can be easily possible in our daily lives. Stringent carbohydrate restriction is not necessary at all. Every day you may often feel "I'm a little bit hungry" when you are concentrating on your work. In these cases, the concentrations of ketone bodies are within these ranges (0.2–0.5 mM) [Newport (Alzheimer's disease: what if there was a cure?: The story of ketones. Basic Health Publications, 2011)].

7.1 How Do We Protect the Brain?

Why did Mary T. Newport's husband recover from Alzheimer's disease? [1]. Let us look into the issue of how small ketogenic (a small increase, 0.2–0.5 mM, in ketone body concentration) has various physiological effects on the brain [1, 2].

To this goal, we must look at the following three properties of ketone bodies:

(1) They pass freely through the blood-brain barrier.
(2) They act directly on the mitochondria.
(3) They can rapidly exert physiological effects.

I must apologize for repeating this so many times, but I think it is important that people should know that small ketogenic can ameliorate the declining cognitive functions at the onset of dementia. It is my honor that some people who read this book can avoid dementia and the progression of dementia.

7.1.1 Protective Effects of Foods Containing Ketone Bodies

Well, let us look into the experiment by using a mouse Alzheimer's model, as shown in Fig. 7.1.

A. Normal food (carbohydrate:protein:fat = 1:1:1)
B. Carbohydrate-rich food (carbohydrate:protein:fat = 7:2:1)
C. Ketone body-containing food (normal food + ketone body 0.5 g/kg)

The paper compared cognitive functions between Groups A, B, and C. Although cognitive functions declined each day in Groups A and B, in Group C (ketone body-containing food) the decline was close to being suppressed and cognitive function augmented. This drastic change is caused by oral administration of the ketone bodies targeting the hippocampus and eliminating energy shortage. The supply of ketone bodies (Group C) is certainly more effective than that of glucose (Group B)

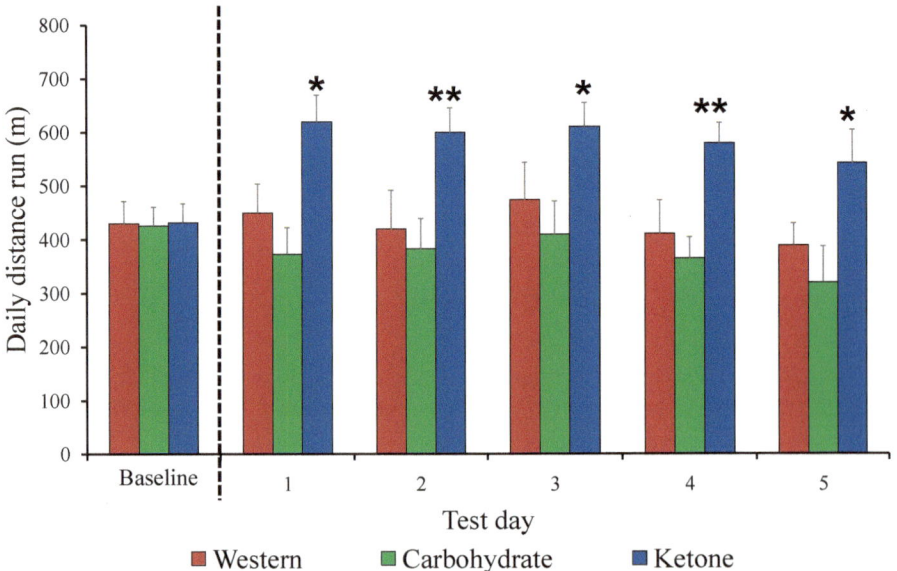

Fig. 7.1 Ketone bodies enhance cognition in mice. Just a ketone body diet can rescue the brain from loss of cognition [3]. Note that this occurs within a single day and that a carbohydrate-rich diet does not induce this effect. Their ketone bodies rescue the early onset of AD diseases. This image is an adapted form of [3]

in terms of the prevention of dementia. However, extensive clinical data have confirmed these results in humans. As ketone bodies can pass freely through the blood-brain barrier and target directly the mitochondria of neurons, this is a unique secondary energy source of the brain. Humans and mice share the same conclusions in terms of prevention against dementia.

7.2   Synthesis of Ketone Bodies

7.2.1   Inhibition of Ketone Body Synthesis by Insulin Spikes

As shown in Fig. 7.2, ketone bodies are synthesized from fatty acids by the complex pathway in the mitochondria of astrocytes and hepatocytes. First, fatty acids are broken into a basic chemical part, known as "acetyl CoA," which is used as a building block for the synthesis of ketone bodies.

Acetyl CoA is a fundamental element in biochemical pathways, acting as a building block. Hepatic cells and astrocytes synthesize ketone bodies from this building block. In this synthesis, HMGCS2 plays a central role, as shown in Fig. 7.2. Insulin spikes potently inhibit HMGCS2 expression. This result is highly important. This is because HMGCS2 is shut out by frequent insulin spikes when excessive carbohydrates are consumed during the meals of breakfast, lunch, and dinner. In this case, the ketone body concentration is 0.1 mM. Most people in the USA and Japan live with this concentration of ketone bodies [5, 6].

Only eight cases of HMGCS2 deficiency are known worldwide. This fact must be related to the importance of the ketone body shuttle in the brain because there are only eight cases in a population of almost 8 billion. Ketone bodies produced by

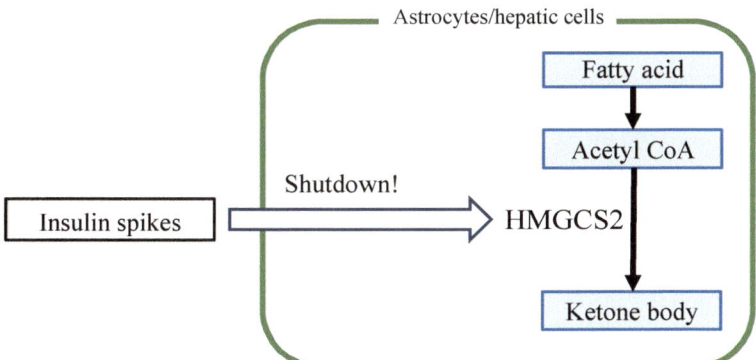

Fig. 7.2 Synthesis of ketone bodies [4]. Ketone bodies are synthesized from fatty acids by HMGCS2, a rate-limiting enzyme in the mitochondria in astrocytes and hepatic cells. This critical enzyme is strictly shut down by insulin spikes. Therefore, excessive insulin spikes should be avoided to increase ketone body concentrations. This is the first action to take

HMGCS2 are highly essential for the survival of *Homo sapiens* under prolonged starvation. *Homo sapiens* have been exposed to severe hunger since starting to stand in East Africa, and HMGCS2 deficiency must be lethal under hunger. Since *Homo sapiens* evolved from other primitive species with little chance of survival without ketone body synthesis, very few individuals with HMGCS2 deficiency exist at present [5, 6].

7.2.2   Flexibility of the Human Brain

Therefore, patients with HMGCS2 deficiency must be careful not to get hypoglycemia in their daily lives. However, they can grow up with a healthy brain if they shorten the intervals between meals. But why do they grow up normally? Why do individuals with HMGCS2 deficiency, who cannot synthesize ketone bodies, allow the brain to grow large? One possible answer is that ketone bodies can be supplied from the mother to the embryo through the placenta and another possibility is that glucose can replace ketone bodies when excessive carbohydrate is supplied during development after birth. These facts are a good illustration of the flexibility of the human brain. Humans can survive for at least 2 months because of the shift from a glucose-based metabolism to a ketone-body-based metabolism [7]. The brain can function on the supply of ketone bodies with a minimal supply of carbohydrates. Similarly, since the brain can grow only on glucose, these persons can develop healthily [8].

7.3   Increase in Brain Blood Flow

7.3.1   Increased Brain Blood Flow by Ketone Bodies

One of the main physiological alternations of dementia and depression is the reduction of brain blood flow. A deficiency of nutrients and oxygen to the brain due to a reduction in brain blood flow leads to an energy shortage for pyramidal neurons in the hippocampus. Energy shortage, triggering dementia, is caused by the reduction of brain blood flow. Through these backgrounds, an increase in the brain blood flow may restore cognitive functional decline during the onset of dementia [9, 10].

"The augmentation of brain blood flow by ketone body injection" was reported by an Australian research group in 2018 (Fig. 7.3). Ketone body concentrations increased to 5.5 mM (55-fold greater than the normal concentration) by injection (4 h) of 75 g of ketone bodies to nine volunteers and brain blood flow increased by 30%, as measured by positron emission tomography, suggesting that ketone bodies increase brain blood flow and eliminate the energy shortage in pyramidal neurons of the hippocampus [11].

Fig. 7.3 Ketone body increases brain blood flow, blood ketone body concentration 5.5 mM, and brain blood flow 30% increase in the persons injected with ketone body ($n = 9$) [11]. Ketone bodies increase the metabolic rate of the brain as well as brain blood flow, suggesting that ketone bodies should activate basic brain functions

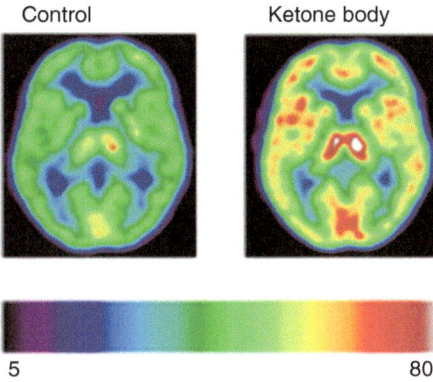

The authors used a direct injection of ketone bodies into volunteers, but it remains unknown for how long the ketone body concentration increased. Thus, it is hard to generalize the results. At least, this paper suggests the possibility that ketone bodies increase brain blood flow and cognitive function. We hope that a direct injection of concentrated ketone bodies may be used as a therapeutic in dementia. This may open new horizons for ketone body research.

7.4   Specific Receptors for Ketone Bodies

7.4.1   Three Special Actions of Ketone Bodies

Ketone bodies are targeted to mitochondria and are used as an energy source to produce the energy substrate ATP as described in Chap. 5. This action is almost the same as other organic acids, such as pyruvate. However, ketone bodies have a special function that other organic acids do not have. Therefore, this topic will become more interesting when you discover the hidden features of ketone bodies, their extraordinary performance, and their amazing functions. However, I will try to provide a rational scientific explanation. The three topics mentioned here are shown in Fig. 7.4.

Since 3-hydroxybutyrate has specific receptors, it can exert various physiological actions in the brain. Cells with special abilities are awaiting ketone bodies to begin their special jobs. A receptor is a protein on the cell membrane that is always waiting to be switched on. As an analogy, you can imagine a seat in an amusement park that starts to move when the park's customers get on it. When the ketone bodies reach the seat, the machine starts to move around and starts running on curved rails. This is amazing because other organic acids do not have this ability [12].

Here, I introduce three receptors for ketone bodies.

(1) HCAR2 (a receptor on the cell membrane) is involved in various physiological functions, such as anti-inflammation and maintenance of memory [13–15].

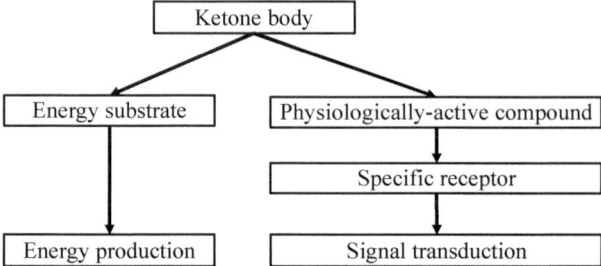

Fig. 7.4 3-Hydroxyutyrate (3HB: the most abundant compound of ketone bodies) activates specific receptors as a physiologically active compound and induces signal transduction pathway as well as functions as an energy substrate for mitochondrial metabolism [12]. Therefore, 3HB is highly distinctive from other organic acids such as lactate, which works just as an energy substrate

(2) GPR43 (a receptor on the cell membrane) is involved in the consumption of fat [16].
(3) HDAC (a receptor in the nucleus) induces the production of antioxidant enzymes and helps to protect the brain [17].

By activating these receptors, ketone bodies can exert various physiological actions. Ketone bodies can exert these special actions owing to specific receptors. In contrast, other organic acids are just used as an energy source for mitochondria. It is like being a normal human but being friends with many wizards and being able to make various requests. Wouldn't that be exciting?

7.5   HCAR2

7.5.1   HCAR2 Activation by Small Ketogenic

First, the authors examined the relationship between 3-hydroxybutyrate and the HCAR2 protein (Fig. 7.5). At a concentration of 0.5 mM, ketone bodies can activate HCAR2 to 30% and the activation is significant. This is a very important result, indicating that a slight increase in ketone body concentration can exert various health-beneficial effects on the body. Thus, small ketogenic (ketone body concentrations of 0.2–0.5 mM) have physiological effects on the brain, especially on pyramidal neurons (short-term memory) as described in Chap. 6. The presence of HCAR2 enables the small ketogenic to power the brain. A vast number of reports and studies [13–15] have reported that these concentrations of ketone bodies exert anti-inflammatory effects on the brain. Thus, the small ketogenic could rescue the husband of Mary T. Newport from the decline of cognitive function [1].

This is why the author assumes that the contentious small ketogenic (0.2–0.5 mM ketone body concentration) empowers the brain and induces sufficient health effects. In addition, a ketone body concentration of 1 mM can induce 75% activation of

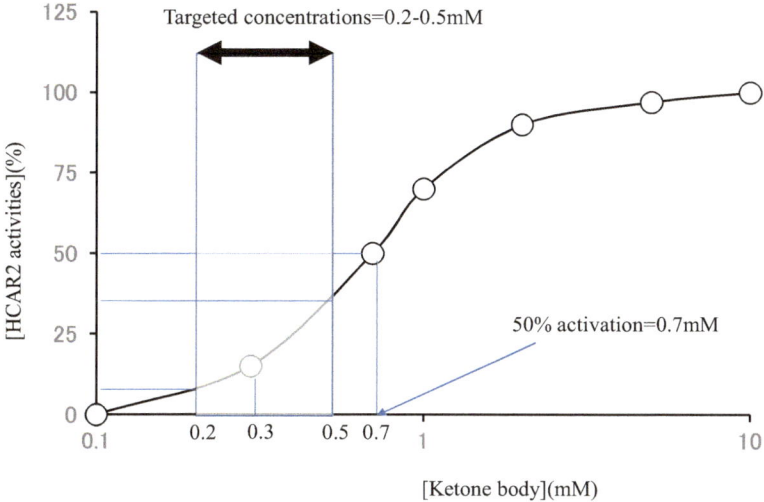

Fig. 7.5 HCAR2 activation by 3HB [13]. Note that 0.2–0.5 mM 3HB (small ketogenic) can significantly activate the HCAR2 receptor. These concentrations of 3HB can have health effects. These concentrations of ketone bodies may induce various health effects as well as extend longevity

HCAR2 and those of 0.2–0.5 mM can significantly activate the HCAR2. However, it is not so easy to maintain 1 mM ketone bodies, because most carbohydrates should be eliminated from daily foods. Small ketogenic (0.2–0.5 mM ketone bodies) can induce various health effects. It is more important that you are not required to make concerted efforts to eliminate carbohydrates; instead, you should be continuously careful of extending the interval between meals and taking light exercise. You may experience small ketogenic every day when you feel mild hunger before dinner. Furthermore, small ketogenic is easy to continue. Only a small effort is required to practice small ketogenic, which empowers the brain.

7.6 GPR43

7.6.1 Fat Burning by Ketone Bodies Mediated by GPR43

3-Hydroxybutyrate has another receptor, GPR43. GPR43 is highly expressed in adipocytes or skeletal muscle cells, but its ligand was not known until recently. In 2019, groups at Tokyo University of Agriculture and Technology identified ketone bodies as its ligand (Fig. 7.6).

In addition, ketone bodies no longer induce fat burning in mice with the GPR43 gene deleted, suggesting that GPR43 is essential for fat burning. What does this result indicate? GPR43 may mediate the effects of fat burning by ketone bodies and has proven to be a receptor for ketone bodies. Before the paper was published, it was

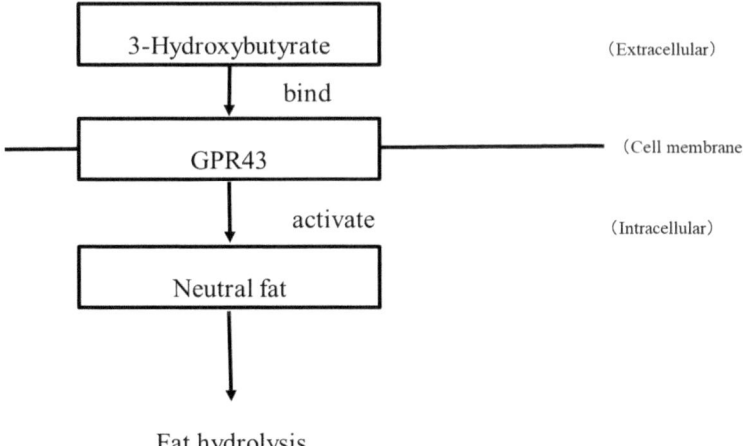

Fig. 7.6 GPR43 activation by 3HB [16]. 3HB binds to GPR43 and induces fat hydrolysis. 3HB is supposed to decrease neutral fat by enhancing fat hydrolysis and burning. GPR43 activation mediates these effects of 3HB

not known why ketone bodies induce fat-burning, although many papers have reported the fat-burning effects of ketone bodies. The paper describes the following sequence of events leading to weight loss by ketone bodies [16]:

1. Binding of ketone bodies to GPR43
2. Signal transduction of GPR43
3. Enhanced fat hydrolysis

   This paper describes the physiological reasons why ketone bodies (small ketogenic) are effective for the prevention of obesity. It turned out to be ketone bodies that acted as the fat-burning machine when the spotlight was focused on their mysterious behavior. In addition, this is why stringent carbohydrate restriction can decrease body weight within a very short period.

7.7 HDAC

7.7.1 Induction of Antioxidant Enzymes by Ketone Bodies

Another protein that binds to 3-hydroxybutyrate is histone deacetylase (HDAC), as shown in Fig. 7.7. The ketone bodies facilitate the induction of 3-hydroxybutyrate, which can pass through the monocarboxylic acid transporter (MCT), enter neurons, and bind to HDAC protein in nuclei. Finally, ketone bodies induce antioxidant enzymes and confer resistance to oxidative stress. Although HDAC usually inhibits the expression of antioxidant enzymes, ketone bodies increase the expression of

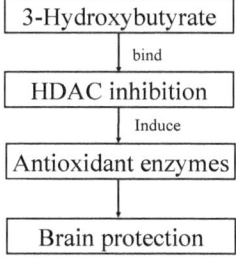

Fig. 7.7 Antioxidant activities of ketone body [17]. 3HB can induce a set of antioxidant enzymes through inhibition of HDAC and protect neurons against oxidative stress. 3HB may activate the transcription of antioxidant enzymes. This transcriptional activation is initiated by the shut-down of HDAC by 3HB

these enzymes by suppressing HDAC. Antioxidant enzymes remove reactive oxygen species and protect neurons from oxidative stress. This paper [17] is highly valued because it identified the molecular mechanism of neuroprotection by ketone bodies. (For example, what if the characters in costumes came to the people who had been waiting for a long time in front of the ticket office and handed out a free pass to everyone? The line disappears in an instant and the atmosphere brightens. This is a comparison of the beneficial effect of eliminating the suppression of antioxidant enzymes.)

In addition, many papers have reported that ketone bodies are effective against Alzheimer's and Parkinson's diseases and depression. In particular, ketone bodies may rapidly restore cognitive functions in patients with Alzheimer's disease as described in the book of Mary T. Newport [1]. Please refer to the references at the end of this book. Ketone bodies have various favorable effects on other organs than the brain, as they have several receptors and are used as an energy source. I hope that readers have become interested in physiology, pharmacology, and brain science.

How did you like this chapter? I feel honored that you are interested in the role of ketone bodies and glucose in the brain. This is of strong practical interest, as this issue is closely related to your brain, body, and aging. In addition, you can achieve small ketogenic today. You can modify your eating habits by yourself. You can be sure that the small ketogenic may have a great contribution to your daily life. Start by taking care of your eating habits. Sometimes it is good to visit an amusement park and have a good day without being mindful of your eating habits.

References

1. Newport MT. Alzheimer's disease: what if there was a cure?: The story of ketones. Basic Health Publications; 2011.
2. Hertz L, Chen Y, Waagepetersen HS. Effects of ketone bodies in Alzheimer's disease in relation to neural hypometabolism, β-amyloid toxicity, and astrocyte function. J Neurochem. 2015;134(1):7–20.

3. Murray AJ, Knight NS, Cole MA, Cochlin LE, Carter E, Tchabanenko K, Pichulik T, Gulston MK, Atherton HJ, Schroeder MA, Deacon RM, Kashiwaya Y, King MT, Pawlosky R, Rawlins JN, Tyler DJ, Griffin JL, Robertson J, Veech RL, Clarke K. Novel ketone diet enhances physical and cognitive performance. FASEB J. 2016;30(12):4021–32.
4. Nakamura MT, Yudell BE, Loor JJ. Regulation of energy metabolism by long-chain fatty acids. Prog Lipid Res. 2014;53:124–44.
5. Rescigno T, Capasso A, Tecce MF. Involvement of nutrients and nutritional mediators in mitochondrial 3-hydroxy-3-methylglutaryl-CoA synthase gene expression. J Cell Physiol. 2018;233(4):3306–14.
6. Sikder K, Shukla SK, Patel N, Singh H, Rafiq K. High fat diet upregulates fatty acid oxidation and ketogenesis via intervention of PPAR-γ. Cell Physiol Biochem. 2018;48(3):1317–31.
7. Cahill GF Jr. Fuel metabolism in starvation. Annu Rev Nutr. 2006;26:1–22.
8. Hasselbalch SG, Knudsen GM, Jakobsen J, Hageman LP, Holm S, Paulson OB. Brain metabolism during short-term starvation in humans. J Cereb Blood Flow Metab. 1994;14(1):125–31.
9. Takano H, Motohashi N, Uema T, Ogawa K, Ohnishi T, Nishikawa M, Kashima H, Matsuda H. Changes in regional cerebral blood flow during acute electroconvulsive therapy in patients with depression: positron emission tomographic study. Br J Psychiatry. 2007;190:63–8.
10. Nordberg A, Rinne J, Kadir A, et al. The use of PET in Alzheimer's disease. Nat Rev Neurol. 2010;6:78–87.
11. Svart M, Gormsen LC, Hansen J, Zeidler D, Gejl M, et al. Regional cerebral effects of ketone body infusion with 3-hydroxybutyrate in humans: reduced glucose uptake, unchanged oxygen consumption and increased blood flow by positron emission tomography. A randomized, controlled trial. PLoS One. 2018;13(2):e0190556.
12. Newman JC, Verdin E. Ketone bodies as signaling metabolites. Trends Endocrinol Metab. 2014;25(1):42–52.
13. Taggart AK, Kero J, Gan X, Cai TQ, Cheng K, Ippolito M, Ren N, Kaplan R, Wu K, Wu TJ, Jin L, Liaw C, Chen R, Richman J, Connolly D, Offermanns S, Wright SD, Waters MG. (D)-beta-Hydroxybutyrate inhibits adipocyte lipolysis via the nicotinic acid receptor PUMA-G. J Biol Chem. 2005;280(29):26649–52.
14. Rahman M, Muhammad S, Khan MA, Chen H, Ridder DA, Müller-Fielitz H, Pokorná B, Vollbrandt T, Stölting I, Nadrowitz R, Okun JG, Offermanns S, Schwaninger M. The β-hydroxybutyrate receptor HCA2 activates a neuroprotective subset of macrophages. Nat Commun. 2014;5:3944.
15. Kovács Z, D'Agostino DP, Diamond D, Kindy MS, Rogers C, Ari C. Therapeutic potential of exogenous ketone supplement induced ketosis in the treatment of psychiatric disorders: review of current literature. Front Psychiatry. 2019;10:363.
16. Miyamoto J, Ohue-Kitano R, Mukouyama H, Nishida A, Watanabe K, Igarashi M, Irie J, Tsujimoto G, Satoh-Asahara N, Itoh H, Kimura I. Ketone body receptor GPR43 regulates lipid metabolism under ketogenic conditions. Proc Natl Acad Sci U S A. 2019;116(47):23813–21.
17. Shimazu T, Hirschey MD, Newman J, He W, Shirakawa K, Le Moan N, Grueter CA, Lim H, Saunders LR, Stevens RD, Newgard CB, Farese RV Jr, de Cabo R, Ulrich S, Akassoglou K, Verdin E. Suppression of oxidative stress by β-hydroxybutyrate, an endogenous histone deacetylase inhibitor. Science. 2013;339(6116):211–4.

Chapter 8
Insulin and Ketone Bodies

Abstract This chapter focuses on insulin in terms of regulation of ketone body synthesis. Since excessive insulin spikes completely shut down the synthesis, we first have to hold insulin spikes to allow ketone body concentrations to increase to a small extent. It was 7 million years ago that human beings appeared in the east Africa. Since they had spent much of their time starved, their body had not been designed so that they could have as many carbohydrates as they liked. Thus, insulin has been very sober and unobtrusive. Human beings have hardly experienced the situation that insulin is so frequently used that the effect of insulin becomes weak. Even in the modern ages, diabetes has been regarded as a luxury disease although it is well-known. Diabetes was considered a disease that rich people could get. It is since the half of the twentieth century diabetes has become a spreading disease that affects the health of modern human beings.

8.1 Insulin Causes Obesity

8.1.1 Insulin as an Obesity Hormone

The functions of insulin can be explained one by one. We will progress believing that there is great interest in this topic. As it is such an important hormone, many researchers have studied insulin. By the early 1990s, the following mechanisms became clear in terms of the progression of diabetes:

1. Insulin secretion
2. Insulin receptor
3. Insulin signal transduction

Because ketone bodies and insulin have a front-and-back relation to each other, the function of insulin can be explained in terms of ketone bodies [1, 2].

The relationship between insulin and ketone bodies can be imagined as a seesaw in the park: when the insulin concentration goes up, the ketone body concentration goes down and when insulin concentration goes down, the ketone concentration goes up. The usual concentrations of ketone bodies are just 0.1 mM because frequent insulin spikes inhibit the enzymes responsible for ketone body synthesis. Furthermore, the cause of insulin spikes is excessive intake of carbohydrates. Here, the basic functions of insulin can be described by the following three actions [3, 4]:

1. Decreased blood glucose
2. Inhibition of ketone body synthesis
3. Synthesis of fat

That is, as insulin spikes, the following are induced:

1. Decreased blood glucose
2. Decreased ketone body concentration
3. Increased fat synthesis

In this background, insulin is regarded as the obesity hormone. There is a scientific basis for obesity originating from the habit of excessive carbohydrates. In contrast, there is a scientific reason why the restriction of carbohydrates leads to a loss in body weight. This leads to the opinion that there is a complete dependence on insulin on whether humans become obese or lose weight. If I can say this without fear of misunderstanding, carbohydrate restriction is performed to decrease the frequency of insulin spikes. The reason why body weight is going to decrease is the effect of insulin, the most powerful obesity hormone. If you want to lose weight within the shortest time, carbohydrate restriction may be the best choice. It is not an exaggeration to say that the way to lose weight is to limit carbohydrates. Only fasting is a more effective method of losing weight than carbohydrate restriction.

Insulin is proposed as an obesity hormone. Insulin may be condemned as the cause of obesity and aging. In this background, the extreme opinion may be that insulin spikes should be eliminated by strict carbohydrate restriction. However, we should pause here and consider this issue [3, 4].

8.2   The Small Ketogenic in Terms of Insulin and Ketone Bodies

8.2.1   Insulin Is an Essential Hormone

However, the human body is not simple. Humans cannot live a healthy and antiaging life by the stringent restrictions of carbohydrates. If you do not eat carbohydrates and keep away from sugar, you will not be able to live well without aging. The brain does not feel happy, and this cannot lead to antiaging, as shown by the study of blue zones all over the world (the area with many individuals with healthy longevity) and by the study of Japanese villages of longevity. It is noted that the residents cannot attain healthy longevity due to giving up bread, rice, noodles, potatoes, and fruits [5, 6].

Therefore, it may be a wise choice to avoid a strict restriction on carbohydrates unless it is medically required (i.e., for therapeutic use). In conclusion, performing a moderate (not stringent) restriction of carbohydrates is a good, antiaging diet. Do not take too much care for the reduction of carbohydrates and avoid excess consumption of carbohydrates. In other words, insulin should be used carefully, but without overuse, partly because insulin is installed as a necessary hormone and partly because insulin has been acquired through the long and tough starvation of human history.

The most important conclusion is that it may be harmful to run to the two extreme concepts: one is complete vegetarianism; the other is strict restrictions of carbohydrates. Simply speaking, one should eat carbohydrates, but not too much. For example, 30% of carbohydrates of total energy are too little, and 70% are too much. Keeping out stringent carbohydrate restrictions allows you to enjoy freshly cooked white rice with egg over rice, baked crepes on a holiday morning, and sandwich ham and lettuce; buy a fruity cake on a tiring day. Enjoying such eating habits may brighten your life [7].

(Although this is a complete digression, Hermann Hesse in Germany wrote a novel. A famous Indian practitioner finally got a job that he could enjoy every day, while he tried every rigorous practice and enlightenment of Buddha. I hope this is a good analogy.)

8.2.2   Proposal of the Small Ketogenic Diet

The estimated proper restriction of carbohydrates is around 0.2–0.5 mM ketone bodies as described in *the HCAR*. Please refer to my standard diet to keep the ketone body concentrations at 0.2–0.5 mM (Fig. 8.1). According to my experience, keeping this eating habit for 2–3 weeks allows you to keep the range of ketone body concentrations. The essence of this eating habit has two important points:

1. Carbohydrates of breakfast are replaced by fat, such as butter.
2. Carbohydrates for dinner are free.

By keeping this eating habit, I am sure that the brain becomes bright.

To enable the hybrid energy system of glucose and ketone bodies, what is the key? The answer is the regulation of insulin spikes. Insulin and ketone bodies are like two sides of the same coin. When insulin spikes occur frequently, the concentration of ketone bodies does not increase from 0.1 mM. When insulin spikes are prevented, these begin to increase. It is insulin that always makes difficult adjustments, such as flipping the coin. Insulin always performs the important job of distributing energy supply to the body and brain. I hope I could have fully restored the honor of insulin. That is why ketone body synthesis is under complete control (inhibition) of insulin because insulin is a potent inhibitor of HMGCS2, a critical enzyme for ketone body synthesis. When insulin spikes are prevented, the synthesis of ketone bodies starts in the liver and astrocytes of the brain. In contrast, when insulin

Proposed Small Ketogenic Diet

Carbohydrates occupy 40-50% of total calories per day.

• Breakfast = butter coffee

• Lunch = 2boiled eggs + 2bananas

• Dinner = meal without limitation

Lunch

Fig. 8.1 My sample of the small ketogenic diet during a week. Note that 40–60% of carbohydrate diet can keep ketone body concentrations within 0.2–0.5 mM, which might be able to extend healthy longevity. I do not recommend my sample of small ketogenic. I introduce my sample just because it may become a reference for you to design your small ketogenic diet by yourself. This diet plan is produced just for me to hold insulin spikes and keep my body within 0.2–0.5 mM ketone body concentrations. But this description is opened just for you to grasp the essence of the small ketogenic by observing my sample. The essence of my small ketogenic diet has the following four points. (1) Insulin spikes are inhibited as far as I can in the daytime. (2) Insulin spikes in the morning should be zero. (3) Daily appetite is satisfied by the free intake of carbohydrates in the dinner. (4) Feeling hunger can be regulated by decreasing and increasing the number of bananas. 3HB concentrations were revealed to be within 0.2–0.5 mM by the small ketogenic in Table 8.1. Please note that this is just my sample. Therefore, this result cannot be generalized. This data is shown here to present just a reference for readers of the book

Table 8.1 Regulated 3HB concentrations within a week

	3HB concentrations (mM)
Monday	0.2
Tuesday	0.5
Wednesday	0.4
Thursday	0.3
Friday	0.5

3HB concentrations were measured at 15:00 by use of Precision III (Abbott). Note that the concentrations of ketone bodies are within the small ketogenic range (0.2–0.5 mM)

spikes frequently occur, ketone bodies cannot increase. The concept is one of the essential concepts for "small ketogenic" [5, 6]. Here, I introduce my proposed method to keep ketone body concentrations within 0.2–0.5 mM and this is very easy to try.

According to my experiences of living a small ketogenic life during these years, the author assumes that the elderly individuals in *The Blue Zones* by Dan Buettner [8] and in the village of healthy longevity in old Japan examined by Shoji Kondo [9]

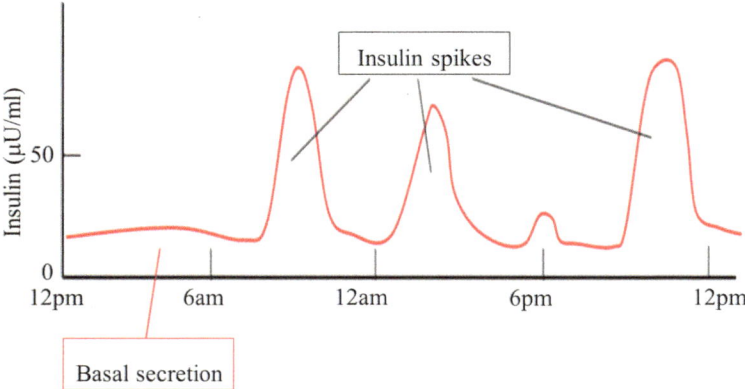

Fig. 8.2 Insulin spike and basal secretion (https://okada-dmcl.jp/blog/post-79). Note that insulin levels have two components, insulin spikes and basal secretion. Insulin spikes are due to glucose spikes and basal secretion is not affected. Basal secretion is essential for the survival of humans. Insulin spikes are induced by the intake of carbohydrates at breakfast, lunch, and dinner. As far as basal secretion is concerned, ketoacidosis cannot occur

may have small ketogenic (0.2–0.5 mM ketone bodies) as mentioned in Fig. 8.1. If you wish to achieve healthy aging or an empowered brain (rather than therapeutic action against specific diseases), you are not required to perform stringent carbohydrate restriction. Rather, this approach may cause various risks as described in Fig. 9.3. The effort of continuing small ketogenic is of much more importance to your daily life. This is because small ketogenic can empower the brain and may extend healthy longevity.

8.3   Insulin Spikes

8.3.1   Basal Secretion and Insulin Spikes

As shown in Fig. 8.2, there are two components to insulin concentrations. One is insulin spikes; the other is basal secretions. As basal secretions are highly necessary for survival and growth, it is advisable not to alter this component.

When the basal secretion of insulin is working, ketone bodies cannot increase without limitation (Table 8.2). In other words, "ketoacidosis" may be possible just when the basal secretion is not maintained. That is, a healthy person never has high concentrations of ketone bodies (over 10 mM), even if he continues a strong restriction of carbohydrates. This situation is known as "physiological ketosis." Thus, the distinction between "ketoacidosis" and "physiological ketosis" is the basal secretion of insulin. In contrast, when the basal secretion of inulin is destroyed (type 1 diabetes), ketone body concentrations can increase without physiological limitations. This is "ketoacidosis." In this case, blood may become acidic (severe acidosis) [10].

Table 8.2 Differences between ketoacidosis and physiological ketosis

	Basal secretion of insulin	Ketone body concentrations
Ketoacidosis	Not working	Often over 10 mM
Physiological ketosis	Well working	Usually, less 10 mM

Note that the destruction of basal secretion of insulin leads to ketoacidosis. Just high concentrations of ketone bodies, for example >5 mM, are not defined as "keto-acidosis" as far as basal secretion of insulin is working. Ketoacidosis is a serious complication of diabetes. The condition develops when the body can't produce enough insulin. Insulin plays a key role in helping sugar—a major source of energy for muscles and other tissues—enter cells in the body. Without enough insulin, the body begins to break down fat as fuel. This causes a buildup of acids in the bloodstream called ketones. If it's left untreated, the buildup can lead to diabetic ketoacidosis. If you have diabetes or you're at risk of diabetes, learn the warning signs of diabetic ketoacidosis and when to seek emergency care.

Insulin spikes are caused by glucose spikes by eating carbohydrates in the morning, noon, and evening meals. When insulin spikes occur frequently, ketone body concentrations remain at around 0.1 mM. If you try to remove all carbohydrates from your meals, insulin spikes will be shut off. Since the inhibition of ketone body synthesis by insulin spikes is gradually decreasing, the concentrations of ketone bodies start to increase. Ketone bodies do not just increase after carbohydrate restriction. The influence of insulin spikes is huge and lasts for a long time. Thus, it may take some time to start to increase the ketone body concentration [10].

Moreover, the stringent restriction of carbohydrates may have highly effective therapeutic actions against epilepsy, severe diabetes, and obesity. Blood glucose and body weight may return to normal levels within several months. Obesity and diabetes may fade out. High concentrations of ketone bodies achieved by a ketogenic diet are highly effective against repeated epileptic seizures. Recently, the strict restriction of carbohydrates has been highly effective for the treatment of diabetes and obesity. In contrast, the strict restriction of carbohydrates may increase the risk of the toxic amino acid, homocysteine. Thus, stringent carbohydrate restriction should be limited in medical use and is not recommended for a healthy person to keep for a long time. If you are a healthy person, looser restrictions should be much better.

8.4 Delay of Aging

8.4.1 Small Ketogenic as a Countermeasure to Aging

Let us consider possible countermeasures to aging. One can keep a strict restriction on carbohydrates for several months, but can this diet be adhered to for several years? It is almost impossible. It is not necessary to aim for severe carbohydrate restriction. In addition, this radical and extreme diet has a serious risk, for example, methionine-induced cardiac failure by increased animal proteins [7]. However, it is at least required to minimize the negative effect of frequent insulin spikes by adjusting excess carbohydrates. Moderately keeping carbohydrate restriction, so as not to consume excess carbohydrates, maybe the best long-term antiaging measure. This may lead to healthy longevity [7].

By adjusting eating habits to minimize the negative effects of insulin spikes, the ketone body shuttle and lactate shuttle operate together to complement each other, while ketone body synthesis is activated in the liver and brain astrocytes. An abundant energy supply to the brain, enabled by the hybrid system, can strengthen the serotonin system; the brain keeps developing and one can live a hopeful life. Neurons with a serotonin receptor in the frontal lobe are similar to pyramidal neurons in the hippocampus with a glutamate receptor. These neurons have high energy demands associated with their performance (happiness and memory in the frontal lobe and hippocampus, respectively). I believe that easing brain energy problems may be the priority to feel happiness in our daily lives. However, some specific points must be considered:

1. Perform a proper exercise in daily life. Brain blood flow is increased, and neuronal growth is stimulated. Secretion of nerve growth factor (NGF) and brain-derived neurotrophic factor (BDNF) are increased.
2. Extend time intervals between meals. The brain may become free from the influence of insulin spikes and is supplied with energy substrate.
3. Do not consume excess carbohydrates. If carbohydrates are controlled within 40–60% of total calories, insulin spikes are effectively suppressed.

8.5 Insulin and Ketone Bodies

8.5.1 Accumulation of Nutrients by Insulin

Again, let us consider why insulin is an obesity hormone. (You may think that I will condemn insulin despite restoring the honor of insulin. The term "obesity hormone" is not used to criticize insulin.) First, consider the image in Fig. 8.3. This figure shows what events occur when excess carbohydrate is consumed. To simplify the explanation, glucose is set at a starting point [3, 4].

When insulin spikes occur frequently, most of the glucose is not metabolized and used for glycogen synthesis. In skeletal muscle, glucose is not metabolized but is used for fat (triglyceride) synthesis under frequent insulin spikes. In contrast, when insulin spikes are suppressed, the fat breaks down and produces energy. Metabolites can enter muscle mitochondria for fat synthesis, but not for oxidative phosphorylation. This is the physiological basis of insulin as an obesity hormone [1, 2].

Briefly speaking, the human body experiences the following events when insulin spikes frequently occur. Glucose is used for the synthesis of glycogen and fat in the liver and skeletal muscle, respectively, as shown in Fig. 8.3. Finally, the

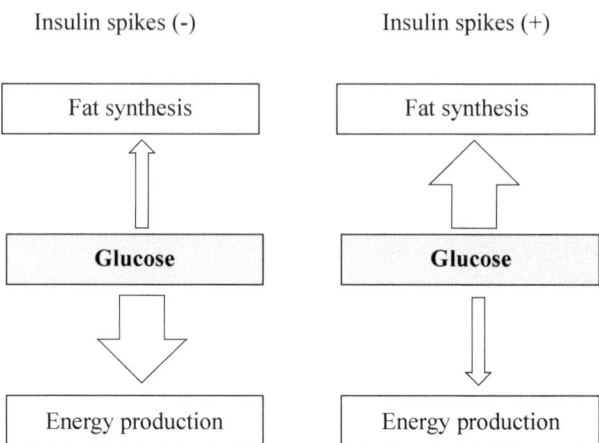

Fig. 8.3 Insulin as a hormone of obesity. Excessive insulin spikes induce obesity in humans. When insulin spikes are frequently evoked, glucose is used for fat synthesis, but not for energy production. In contrast, when insulin spikes are relatively suppressed, glucose is used mainly for energy production

carbohydrates in the foods contribute to an increase in obesity. The basic function of insulin is to store nutrients in the body. This important function can be life-saving. This is why carbohydrates in food are converted to fat. Imagine a human fortuitously finding an apple tree in autumn in the ice age. He has to eat the apples and save a lot of fat in his body until he is full before the season shifts to the winter. In this case, insulin must work hard to save fat; this is essential for the survival of the species.

This is the basic reason why insulin is an obesity hormone. In summary, insulin is doing a hard job. The problem is the eating habits of a modern lifestyle: humans eat too many carbohydrates. In contrast, when insulin spikes are suppressed by the restriction of carbohydrates, nutrients are not saved and are instead used as energy substrates in the body. This is why one can lose weight by restricting carbohydrates for a short period.

8.5.2 Mechanism of Fat-Burning

Now we should consider the relationship between insulin and ketone bodies, as this relationship may have a big impact on the brain energy system. As shown in Fig. 8.4, there are three players in the fat metabolism [3, 4].

1. Adipocytes save large amounts of neutral fat (triglyceride) in the subcutaneous tissue. If you wish to touch it directly, try pinching your tummy with your hands—you will feel a mass of fat cells under the skin. Adipocytes are part of everybody.

Insulin spikes (-)   Insulin spikes (+)

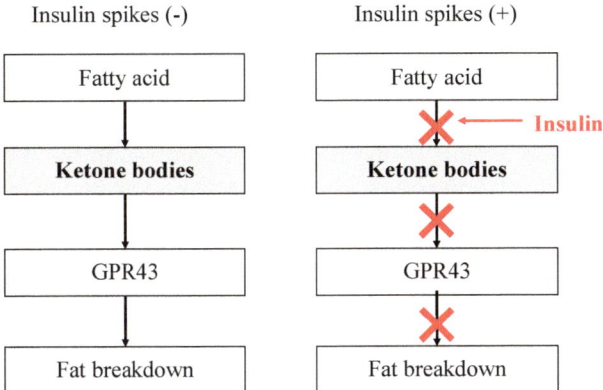

Fig. 8.4 Opposite effects of ketone bodies and insulin. Insulin potently inhibits ketone body synthesis by shutdown of HMGCS2, and ketone bodies induce fat breakdown by acting GPR43 when insulin spikes are relatively suppressed. However, when insulin spikes are frequently evoked, human bodies start to increase weight by inhibiting fat breakdown. Ketone bodies can potently induce fat burning by activating the GPR43 receptor

2. Hepatic cells in the liver synthesize ketone bodies from fatty acids and release them into the blood.
3. Skeletal muscle cells are the largest major consumers of blood glucose and ketone bodies when frequent insulin spikes are suppressed. The cells make the energy substrate ATP by using ketone bodies and glucose.

Have you ever run a marathon? Consider what happens in the body when you run a marathon. The large amount of oxygen taken up in the lungs reaches the skeletal muscles and produces the energy substrate ATP in skeletal muscle cells. ATP is produced initially by the use of glucose. If you run for longer than 20 min, glucose is converted to ketone bodies. Skeletal muscle cells produce the energy substrate ATP by the use of ketone bodies fully oxidized in mitochondria.

The origin of ketone bodies consumed in skeletal muscle cells is a large amount of neutral fat saved in adipocytes under the skin. The neutral fat is hydrolyzed by lipase to produce glycerin and fatty acids, which are released into the blood. Fatty acid binds to proteins and is absorbed by hepatocytes. Ketone bodies are produced here from fatty acids in hepatocytes.

The key enzyme in ketone body synthesis is HMGCS2. Through this enzyme, fatty acids are used to produce ketone bodies in hepatocytes. Therefore, endurance exercise increases fatty acids and ketone bodies in the blood. In addition, as ketone bodies activate GPR43, the hydrolysis of fat is further enhanced in skeletal muscle and under the skin. This is why fat under the skin is hydrolyzed during aerobic exercise. Since ketone bodies can be targeted to the brain and pass freely through the blood-brain barrier and neurons are supplied through astrocytes, the brain becomes clear, and positive feelings are obtained. Thus, aerobic exercise is good in many ways [5, 6].

8.5.3 Shutdown of Ketone Body Synthesis by Insulin Spikes

What happens to the body after one sweet drink and one cold soft drink after the marathon? There is a big glucose spike due to the hydrolysis of sugar. The glucose spike is sensed by beta cells of the islets of Langerhans in the pancreas, which causes a big insulin spike to reduce glucose concentrations in the blood. Through insulin, all cells absorb glucose from the blood and rapidly reduce glucose levels.

In this case, insulin spikes may play the role of a bad guy. Insulin rapidly and completely inhibits the enzymes that are involved in fat hydrolysis. That is, fat hydrolysis is rapidly shut down and all events leading to fat hydrolysis are ruined. Of course, HMGCS2, a key enzyme of ketone body synthesis, is inhibited. Therefore, ketone body synthesis stops, and the concentrations of ketone bodies return to the usual level (0.1 mM). The flow of ketone bodies to the brain is not possible [5, 6].

Therefore, if one eats excess carbohydrates three times every day, ketone bodies cannot work at all. If this continues all year round, it will be troublesome.

In contrast, if you stop eating excess carbohydrates, ketone body synthesis will take place, and fat hydrolysis will be enhanced.

If you understand the statements above, when insulin spikes frequently occur, not only will fat hydrolysis not take place, but also fat synthesis from carbohydrates is promoted. You will unilaterally become obese. It is worse if one does not use the system for consuming fat (ketone bodies). In brief speaking, the frequent daily consumption of excess carbohydrates should be avoided for healthy longevity.

References

1. Banks WA, Owen JB, Erickson MA. Insulin in the brain: there and back again. Pharmacol Ther. 2012;136(1):82–93.
2. Cheng Z, Tseng Y, White MF. Insulin signaling meets mitochondria in metabolism. Trends Endocrinol Metab. 2010;21(10):589–98.
3. Rattigan S, Bussey CT, Ross RM, Richards SM. Obesity, insulin resistance, and capillary recruitment. Microcirculation. 2007;14(4–5):299–309.
4. Kolb H, Kempf K, Röhling M, Martin S. Insulin: too much of a good thing is bad. BMC Med. 2020;18(1):224.
5. Polonsky KS, Sturis J, Van Cauter E. Temporal profiles and clinical significance of pulsatile insulin secretion. Horm Res. 1998;49(3–4):178–84.
6. Cox PJ, Kirk T, Ashmore T, Willerton K, Evans R, Smith A, Murray AJ, Stubbs B, West J, McLure SW, King MT, Dodd MS, Holloway C, Neubauer S, Drawer S, Veech RL, Griffin JL, Clarke K. Nutritional ketosis alters fuel preference and thereby endurance performance in athletes. Cell Metab. 2016;24(2):256–68.
7. Seidelmann SB, Claggett B, Cheng S, Henglin M, Shah A, Steffen LM, Folsom AR, Rimm EB, Willett WC, Solomon SD. Dietary carbohydrate intake and mortality: a prospective cohort study and meta-analysis. Lancet Public Health. 2018;3(9):e419–28.
8. Buettner D. The blue zones. National Geographic; 2012.
9. Kondoh S. Long- and short-lived villages in Japan. Sanroad Publishing; 1991.
10. Cashen K, Petersen T. Diabetic ketoacidosis. Pediatr Rev. 2019;40(8):412–20.

Chapter 9
Insulin Hypothesis

Abstract Why human beings are aging? Many people have been considering this issue for a long time. The reasonable answer is that we have not reached such a conclusion. If we have identified the substance that causes human aging, we would accomplish healthy longevity by avoiding this. However, we have a candidate for an aging substance. In 1993, the paper greatly shocked me that insulin is the real substance for aging in the worm. While considering whether this theory "insulin hypothesis" can be applied to mammalians, the roles of insulin in aging will be discussed here.

9.1 Kenyon and Brenner

9.1.1 Nematodes with Double the Lifespan

In 1993, a surprising paper was published in *Nature* [1]. The field of antiaging research completely changed from before 1993 to after 1993. As the paper required us to alter our common conceptions of insulin, it had a major influence on diabetes research. In addition, because studies of worm longevity had previously not affected human aging research, the paper had a big impact. Here is the abstract of the paper [1] (Fig. 9.1).

> We have found that mutations in the gene daf-2 can cause fertile, active, adult Caenorhabditis elegans hermaphrodites to live more than twice as long as wild type. This lifespan extension, the largest yet reported in any organism, requires the activity of a second gene, daf-16. daf-2 and daf-16 provide entry points into understanding how lifespan can be extended.

Anybody who has ever read scientific papers knows that other papers reporting lifespan extension display only a very small increase. Lifespan extension by 30% is a surprising result. Therefore, "live more than twice as long as wild type" must suggest a special event occurs in the Daf2 mutant, which was very exciting for Kenyon. Indeed, the study of longevity has been considered around the Daf2 protein. This is a wonderful thing. This result was confirmed by researchers all over the world and

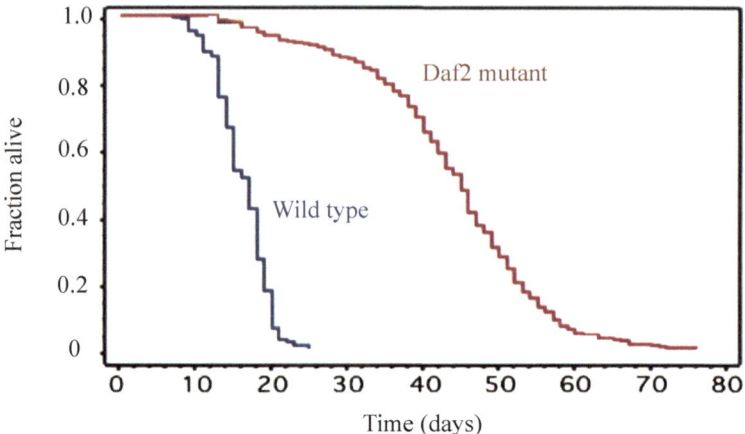

Fig. 9.1 Extended lifespan by Daf2 mutation. Please note that the lifespan of Daf2 mutants is over twice as much as that of wild type. Daf2 is a gene with a short lifespan and therefore Daf2 mutant leads to an extended lifespan [2]. Several years later, Daf2 was revealed to be a gene encoding insulin receptor of the worm

subsequently, many researchers have entered this field, forming a large group. In my estimation, Kenyon will receive the Nobel Prize in Physiology or Medicine within 10 years.

This is a typical example of a "Paradigm shift" in the science [3]. With the emergence of this epoch-making work, common sense among scientists has changed within a very short time. From that time (1993), the center of antiaging research had shifted from the Free Radical Theory of Antiaging (FRTA) proposed by Herman Denham [4] to the insulin hypothesis proposed by Cynthia Kenyon [1]. In other words, most researchers believed that free radicals cause aging up to 1993. After the emergence of the paper [1], many researchers, who had studied the aging of small animals such as worms, flies, and mice became enthusiastic supporters of the insulin hypothesis. However, researchers in human antiaging had not believed the insulin hypothesis up to recently because insulin had been used as a therapeutic against diabetes. Within these several years, the insulin hypothesis has been spreading more and more into the medical field [5, 6]. The issue of whether insulin causes human aging remains to be discussed further.

9.1.2   The Study Was Initiated by Brenner

C. elegans (nematode worm) used for longevity studies is a small animal, about 1 mm in length. Although some of the closely related species are famous as parasites, *C. elegans* is a species that lives independently in the soil, but it is not a parasite. It was in the 1960s that *C. elegans* began to be used as an experimental animal by Sydney Brenner, a famous molecular biologist. Brenner performed Nobel Prize-winning research in the early days of molecular biology in the 1950s and he left the field of bacterial molecular biology, saying that all the important work in molecular biology had been completed. At that time, most of the other researchers did not understand his need to change his topic of study.

Speaking of the situation in Japan, it was in the 1960s that biological research was first influenced by molecular biology. He stated that bacterial molecular biology had ended; he must have outstanding foresight. He said that the next issue was molecular biology, focusing on the development of multicellular organisms and introduced *C. elegans* (worm) into molecular biology techniques and started research into the developmental biology of the worm. It can be said that all studies using *C. elegans* are a result of his thinking. The worm is highly easy to maintain. The culture of the worm does not need the same sterility that is used in cell culture. It is highly easy to expand as the worm can proliferate more and more when supplied only with *E. coli*. In addition, morphological observation is easy. The size (1 mm) is large enough that can be seen under low magnification. It is a valuable feature of developmental biology that everything inside the body is transparent, and the organs can be seen under a microscope. Since the worm has just 1000 cells in total, all cellular differentiation during development was clarified relatively quickly.

Many great works have come out of the study on the worm. The most famous is the study of cell death. The Nobel Prize in Physiology and Medicine was given to Brenner, Horwitz, and Sulston in 2002. From this award, the new work Brenner had probed to confirm his beliefs, which were approved by the Nobel Prize Committee. Incidentally, many researchers found it hard to change the subject of study because they are likely to lose many friends and funding sources. This is a similar situation that you face when you move to a new company in an unfamiliar field after you have quit a company in which your contribution has been highly appreciated.

9.1.3   Reverse Genetics

The worm has another outstanding feature. They have a short lifespan, of just 2 weeks. Therefore, it is suitable for the longevity research of multicellular organisms. For example, because the mouse has a 2-year lifespan, about 3 years are needed to complete a single experiment. In contrast, with the worm, a month is required. Therefore, the worm is the best experimental animal for longevity research. Under these circumstances, epoch-making studies have been continuously produced since the 1980s.

Reverse genetics is an often-used methodology in these fields. Having reverse genetics means there is something called forward genetics. In the nineteenth century, Gregor Johann Mendel, who discovered Mendelian laws, utilized forward genetics, and initiated the genetics.

Forward genetics: assume a gene and explore the expression pattern of the trait.
Reverse genetics: study a mutant, identify the responsible gene, and explore the
    biological mechanisms.

Reverse genetics is mainly used for budding yeast, fission yeast, *Drosophila*, and
nematodes. In this field, a mutant strain of interest can initiate everything. Leland
Hartwell (cell cycle mutants) in yeast, Edward Lewis (somatogenesis mutants) in
flies, and Sydney Brenner (programmed cell death mutants) in worms are famous.
They are all pioneers in their fields and identify many mutants. Their successors
produced great works by using these mutants. All received the Nobel Prize in
Physiology and Medicine. In the field of the worm, the study of mutants of longevity had begun, and are now divided mainly into two groups.

Mutants with long lifespans (e.g., lifespan is twice that of the wild type)
Mutants with short lifespans (e.g., lifespan is one-half of that of the wild type)

9.1.4 Mutants of Short and Long Lifespan

What is a mutant with a long lifespan? Since the disruption of a gene (deletion of a
function) leads to a long lifespan, the gene of interest is involved in the short lifespan of *C. elegans* (worm). In contrast, in the case of a mutant with a short lifespan,
the gene of interest involves a long lifespan. The items are of great importance.

Daf2 mutant has a double lifespan.
Daf16 mutant has a half lifespan.

The two mutants are the most famous mutants of longevity. Then, the original
function of these genes can be rephrased as:

Daf2 is a gene for a short lifespan.
Daf16 is a gene for a long lifespan.

The worm lifespan is shortened by Daf2 and extended by Daf16. This basic
genetic information is accumulated by genetic works using a vast number of longevity mutants and is concentrated in some research groups (e.g., Kenyon's lab) in
1990. Through these genetic works, important knowledge has then been started.
Daf2 and Daf16 are molecules in the same signal transduction pathway. In addition,
they can strongly affect each other. Briefly speaking, the genetic works can be sum-
marized in Fig. 9.2.

The wild type, in which the Daf2 gene for a short lifespan functional, suppresses Daf16, the gene for a long lifespan and only lasts 2 weeks. The Daf2
mutant, in which Daf2 is not functional, allows Daf16 to work and has twice the
lifespan (4 weeks) of the wild type. Cynthia Kenyon discovered this basic genetic
information. She collaborated with Sydney Brenner in the 1980s and had an inde-
pendent lab in the 1990s. The aging research before and after Kenyon's paper

"Insulin hypothesis"

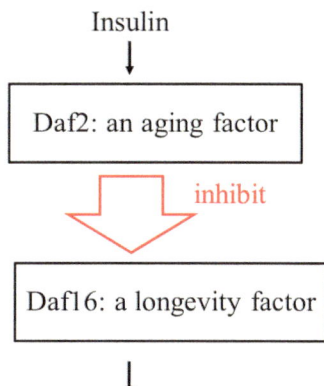

Insulin

Daf2: an aging factor

inhibit

Daf16: a longevity factor

Shortage of lifespan

The human impacts of the Insulin hypothesis are still in discussion.

1. Is insulin an aging hormone?

2. Should insulin be suppressed to delay aging?

Fig. 9.2 The essence of the insulin hypothesis is an inhibition of Daf16, a longevity factor, by Daf2 an aging factor [1, 2, 7]. In the Daf2 mutant, Daf16 is allowed to work to extend the lifespan at least in the worms. The studies on Daf2 and Daf16 have created a breakthrough in the antiaging biology

published in 1993 had become distinctive from each other. After Kenyon's paper in 1993, many epoch-making papers on longevity research were published consecutively in prestigious scientific journals such as *Nature*, *Science*, and *Cell*. Since then, longevity research has been fluctuating around the "insulin hypothesis." The research project on the roles of insulin receptors (Daf2 in the worm) in mammals has become the cutting edge of basic biological science. New findings are published every week during these days. By the way, in the early 1990s, cutting-edge researchers in the field shared the two important issues that should be solved as the next problem.

1. What is the function of Daf2 in humans?
2. What is the function of Daf16 in humans?

9.2 Insulin Hypothesis

9.2.1 Insulin Receptor Is a Candidate for a Gene of Short Lifespan

The paper Kenyon in 1993 was very surprising [1]. This is because this paper suggests that Daf2 may encode an insulin receptor. Daf2 is a homolog of the mammalian insulin receptor gene. In this background, the competition for cloning the Daf2 gene from human cDNA has become intense. Ruvkun was the first to clone the gene and provide direct evidence [7].

Indeed, let us discuss the central insulin hypothesis. However, it is not a fact, but just a hypothesis of aging. Therefore, it is advisable to read forward armed with this knowledge. Insulin holds a very delicate position in medicine and science. According to straightforward investigations and generalization of the results by Kenyon and Ruvkun [5], the following conclusions may be drawn:

1. Insulin is an aging hormone.
2. Insulin should be suppressed to delay aging.

This is called the "insulin hypothesis." Even now, the center of longevity research is the insulin hypothesis. According to the insulin hypothesis, insulin is regarded as a molecular clock of longevity to measure the aging of living creatures.

If the simplest interpretation is allowed here, the following statements are available. The more insulin you use, the older the organisms. To slow aging, it is best to use insulin in small increments. Many scientists think that insulin spikes should be avoided as much as possible.

9.2.2   FOXO3 Is a Gene for Longevity

This was a required part, "a missing link" for the completion of the insulin hypothesis. That is the issue of what Daf16 is in mammals. Ruvkun's and Kenyon's groups reported this in 1997 [7–9]. They clarified that Daf16 is a homolog of FOXO3, a transcriptional factor, which was already famous as one of the candidate longevity genes. A transcriptional factor is a protein that binds to DNA to activate or inhibit the transcription of specific mRNAs. There are many reports that FOXO3 in mice induces various proteins, which are involved in healthy effects. Against this background, FOXO3/Daf16 was quickly accepted as a longevity gene by researchers who studied aging mechanisms using mouse models [7–9].

For example, FOXO3 significantly increases the transcription of HMGCS2 responsible for ketone body synthesis. In contrast, insulin induces the phosphorylation of the FOXO3 protein and the export of FOXO3 from the nucleus to the cytosol. Finally, insulin potently suppresses the health-beneficial effect of FOXO3. In addition, insulin also inhibits HMGCS2, which is responsible for ketone body synthesis. These elucidations were made just after it was discovered that FOXO3 was Daf16 [6–8].

Briefly speaking, the insulin receptor inhibits FOXO3, a potent longevity gene product. Therefore, when insulin spikes are frequently induced, aging is promoted. The key to antiaging is an eating habit that mitigates insulin spikes as much as possible [1, 2, 7].

9.3 Validation of Insulin Hypothesis

9.3.1 Low Levels of IGF-1 in Centenarians

In the insulin hypothesis, many scientists think that Daf2 in the worm should correspond to the mammalian insulin-like growth factor 1 (IGF-1) receptor. Insulin and IGF-1 are like brothers. Thus, it cannot simply be concluded that the cause of aging in mammals is insulin [5–7]. IGF-1 is a hormone similar in molecular structure to insulin that plays an important role in childhood growth and has anabolic effects in adults. During the evolution of mammals, IGF-1 and insulin as well as IGF-1 receptor and insulin receptor have appeared by gene duplication. Therefore, they share the same biological actions in some circumstances and differ in other circumstances.

In the worm, the insulin hypothesis was proposed by the study of Daf2 and Daf16 mutants. The worm, which has only a single hormone, such as insulin, has caused problems by proposing the insulin hypothesis. However, the situation is more complex in mammals. The mammalian genome size is so huge that most of the important genes are duplicated in mammals. Since insulin is an important gene, it has been duplicated, too. Insulin has been duplicated to the following three genes:

Insulin
IGF-1
IGF-2

The debate on which the mammalian gene corresponds to worm insulin was sparked by Kenyon's paper in 1993 [1]. In conclusion, the opinion that the worm Daf2 protein is the mammalian IGF-1 receptor has been supported by many researchers. Consequently, it is suggested that IGF-1 is an aging protein. The results in Table 4.6 show that human aging is highly correlated with the IGF-1 system. Human studies will be required for the conclusions on human aging. The authors performed detailed and extensive blood tests of 100-year-old, their descendants, and people in the control group [5, 6].

9.3.2 IGF-1 Is a Candidate Aging Hormone

This test was performed to identify an inheritable trait of longevity. Centenarians and control individuals were compared in *Comparison 1 in Table 9.1*. Insulin, IGF-1, and IGF-2 were significantly decreased in centenarians, suggesting that these insulin-related hormones are involved in the longevity of humans.

It may be etiologically sufficient, but it is not genetically sufficient. The descendants should be compared. This is because the study aimed to determine if the traits required for longevity were inherited by their descendants. The comparison between the control population and the centenarians' descendants should be

Table 9.1 IGF may be a determinant of longevity [6]

	Control	Centenarians	Offspring of centenarians
IGF-1 (nmol/L)	17	9.3*	14.4**
IGF-2 (nmol/L)	114	72*	134
Insulin (nmol/L)	0.074	0.039*	0.071

Comparison 1: control vs. centenarians (* <0.05)
Comparison 2: control offspring vs. centenarians' offspring (** <0.05)

significantly different (*Comparison 2 in Table 9.1*). In Comparison 2, only IGF-1 produced a significant difference among insulin, IGF-1, and IGF-2. This result suggests that low levels of IGF-1 are required for longevity. In summary, Daf2 in the worm may be equivalent to the mammalian IGF-1 receptor. However, insulin should not be condemned. Rather, the cause of aging in mammals may be IGF-1, although, there are many opinions surrounding the insulin hypothesis. For mammals, the hypothesis is normally described as the "insulin/IGF-1 hypothesis." In mammals, we cannot determine a clear conclusion. However, it seems certain that IGF-1 can function as an aging hormone. Research on aging will accelerate in the future.

9.4   Impact on Carbohydrate Restriction

In particular, those who were particularly sensitive were those who were promoting carbohydrate restriction. Although the Atkins diet became popular in the 1970s and 1980s but faded out, its popularity was restored, and carbohydrate restriction spread all over the world from the USA. Robert Atkins, one of the most famous "carbohydrate restrictionists" all over the world, proposed the Atkins diet of stringent carbohydrate restriction against obesity in his book published in 1972 [10]. He can be appreciated for his influential lay book and for being a pioneer of carbohydrate restriction in such early days. His work is of high value and has made a big difference in the real diet of humans.

His points were the following [10]:

1. Carbohydrates should be replaced by protein and fat (meat).
2. Insulin spikes should be suppressed by decreasing glucose spikes.
3. As a result, induction of ketosis (physiological ketosis) should be aimed.
4. By these eating habits, one can significantly lose body weight.

More simply, a habit of eating excess carbohydrates promotes aging by inducing frequent insulin spikes. Thus, efforts to avoid eating excessive carbohydrates are essential for staying young as long as possible. I may have repeated this too much, but it is a very simple story as the following pathways.

1. Insulin spikes↓ → FOXO3↑ → HMGCS2↑ →Ketone bodies↑
2. Insulin spikes↑ → FOXO3↓ → HMGCS2↓ →Ketone bodies↓

These sequential events may lead to the following simplified conclusions.

1. Stringent carbohydrate restriction → Obesity ↓
2. Excessed carbohydrate intake → Obesity ↑

This pathway is the essence of stringent carbohydrate restriction which Robert Atkins were proposing and many carbohydrate restrictionists are discussing [10]. At least, this diet can lose weight within a short time. However, these events may not lead to healthy longevity in recent works [11].

The Atkins diet was causing the big movement of diet in the USA in the 1970s–early 1980s. Thus, it was not accidental that the stringent carbohydrate restriction spread occurred in the early 2000s. There was a strong scientific background, the "insulin hypothesis." They argued that insulin spikes are produced by glucose spikes and that avoiding glucose spikes is the central issue for antiaging. Even in Japan, some clinicians strongly recommended the carbohydrate restriction and applied it to therapeutics against diabetes in the early 2000s.

9.5   Side Effects of Carbohydrate Restriction

Although they have produced therapeutic effects to some extent, stringent carbohydrate restrictions always with many problems as many papers pointed out as far as antiaging is concerned. Stringent carbohydrate restriction potently increases the hazard ratio as well as an excess of carbohydrates, indicating that around 50% of carbohydrate intake may be the ideal diet, as proposed by meta-analysis as shown in Fig. 9.3. In addition, as far as longevity is concerned, stringent carbohydrate restriction causes the same results as excessive carbohydrate intake.

Therefore, there is a clear conclusion that we have to avoid stringent carbohydrate restrictions as well as an excess carbohydrate diet. Furthermore, it is notable that none of the healthy old in the villages of longevity cut off carbohydrates from their diet. Therefore, we can extract a valuable lesson that stringent carbohydrate restrictions fail to provide positive evidence for human healthy longevity although the diet can lose weight within a short period.

There are three zones of carbohydrate intake in the diet: (A) stringent carbohydrate restriction, (B) targeted carbohydrate intake, and (C) excess carbohydrate intake (Table 9.2). Since both (A) and (C) zones have a high hazard ratio, the (B) zone (carbohydrate 40–60%) is recommended for human longevity. This study shows that both extreme opinions are not justified by etiological research.

Regrettably, extreme opinions are spreading all over the world. They are proposed by the "Fundamentalists of Carbohydrate Restriction," who argue that carbohydrates must be completely cut off to increase ketone bodies. Of course, extreme carbohydrate restriction is highly effective against severe diabetes and refractory epilepsy. However, it is not advisable for long periods. This is because extreme carbohydrate restriction is effective for losing weight in a short period, but it does not contribute to extending healthy longevity in the long run [11]. In addition, there

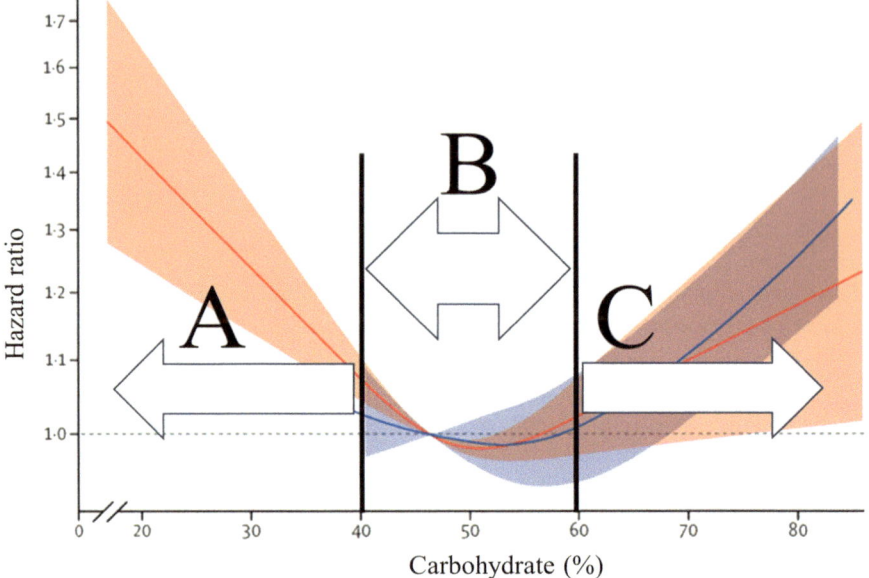

Fig. 9.3 U shape of carbohydrate and hazard ratio. Note that carbohydrate 50% has the lowest hazard ratio and that a lower carbohydrate diet has a higher hazard ratio, suggesting that both an excessive carbohydrate diet and stringent carbohydrate restriction are hazardous to the human body [11]. Hazzard can be interpreted as "likeliness to die." In other words, stringent carbohydrate restrictions as well as excessive carbohydrate intake should be avoided to extend longevity. In addition, the contentious small ketogenic diet is much more effective in extending healthy longevity than stringent carbohydrate restriction

Table 9.2 Intake of carbohydrates and hazard ratio

Zones	A	B	C
Carbohydrates (%)	<40	>40, <60	>60
Diet	Stringent carbohydrate restriction	Targeted carbohydrate intake	Excess carbohydrate intake
Hazard ratio	High	Low	High

is no direct evidence for healthy longevity arising from stringent carbohydrate restriction [11]. These treatments should be performed only for clear therapeutic purposes and limited duration. At least, we do not need complete restrictions in our daily life by a large-scale epidemiological survey. Therefore, it is highly advisable to keep out stringent carbohydrate restrictions.

We do not need to keep extreme carbohydrate restrictions because humans can use glucose and ketone bodies. Little hunger and low ketogenic is possible, even if we have carbohydrates. The problem is not the carbohydrate itself, but excessive carbohydrate intake. Thus, we should keep out of the extreme opinion that carbohydrates should be completely cut off.

We can try to make meal intervals wider than usual. We sometimes try to exercise. We may have other foods to compensate for the small decrease in carbohydrates. I have been measuring the concentrations of glucose and ketone bodies every day for 3 years and I can maintain ketone bodies at 0.2–0.5 mM by these eating habits. Of course, every individual is different, because people have different lifestyle patterns and meals.

References

1. Kenyon C, Chang J, Gensch E, Rudner A, Tabtiang R. A C. elegans mutant that lives twice as long as wild type. Nature. 1993;366(6454):461–4.
2. Libina N, Berman JR, Kenyon C. Tissue-specific activities of C. elegans DAF-16 in the regulation of lifespan. Cell. 2003;115(4):489–502.
3. Kuhn TS. The structure of scientific revolutions. University of Chicago Press; 1962.
4. Harman D. Free radical theory of aging: an update: increasing the functional life span. Ann N Y Acad Sci. 2006;1067:10–21.
5. Brugts MP, van den Beld AW, Hofland LJ, van der Wansem K, van Koetsveld PM, Frystyk J, Lamberts SW, Janssen JA. Low circulating insulin-like growth factor I bioactivity in elderly men is associated with increased mortality. J Clin Endocrinol Metab. 2008;93(7):2515–22.
6. Brugts MP, van Duijn CM, Hofland LJ, Witteman JC, Lamberts SW, Janssen JA. Igf-I bioactivity in an elderly population: relation to insulin sensitivity, insulin levels, and the metabolic syndrome. Diabetes. 2010;59(2):505–8.
7. Kimura KD, Tissenbaum HA, Liu Y, Ruvkun G. daf-2, an insulin receptor-like gene that regulates longevity and diapause in Caenorhabditis elegans. Science. 1997;277(5328):942–6.
8. Hsu AL, Murphy CT, Kenyon C. Regulation of aging and age-related disease by DAF-16 and heat-shock factor. Science. 2003;300(5622):1142–5.
9. Dillin A, Crawford DK, Kenyon C. Timing requirements for insulin/IGF-1 signaling in C. elegans. Science. 2002;298(5594):830–4.
10. Atkins RC. Dr Atkins's new diet revolution: the no-hunger, luxurious weight loss plan that works! Ebury Digital; 2009.
11. Seidelmann SB, Claggett B, Cheng S, Henglin M, Shah A, Steffen LM, Folsom AR, Rimm EB, Willett WC, Solomon SD. Dietary carbohydrate intake and mortality: a prospective cohort study and meta-analysis. Lancet Public Health. 2018;3(9):e419–28.

Chapter 10
Brain and Human Longevity

Abstract How do we get healthy longevity in daily life? To have a possible way for the purpose, we have to look into real eating habits and biochemical tests of the people of healthy longevity. These epidemiological surveys are very important for possible suggestions. Since human has a huge brain, the effects of the brain on longevity are also huge. As far as the brain is concerned, many things remain to be understood by the longevity study by use of worms, flies, and mice. Therefore, insights into the gut-brain axis, blood vessels, and longevity will be done while talking about people who lived long, and those who ended short life. Let's think about how to delay human aging. Since many people are interested in this chapter, I will try my best. Let's proceed while stretching and loosening.

10.1 Blue Zones

10.1.1 Common Features of the Diets in the Villages of Longevity

Dr. Dan Buettner has studied "blue zones" all over the world, areas with a high proportion of healthy and long-lived people (over 100 years old). He studied the following blue zones [1]:

1. Sardinia, Italy
2. Okinawa, Japan
3. Nicoya Peninsula, Costa Rica
4. Ikaria Peninsula, Greek

and said that all of them are vegetarian and have low meat consumption.

According to this report, it is likely that the lifestyle of strictly restricted carbohydrates and an increase in meat to increase ketone bodies may shorten lifespan. Eating meat has the following advantages, it can:

T. Satoh, *Hybrid-Powered Brain*, https://doi.org/10.1007/978-3-031-54150-6_10

1. Avoid too much carbohydrate.
2. Prevent gluten resistance.
3. Provide essential amino acids such as lysine, methionine, and tryptophan.

Therefore, it may not be a bad choice to have much meat for protein intake. This is because excess animal meat has major disadvantages that may cancel out all of the advantages. In many cases, stringent carbohydrate restrictions may fall in the hole of the eating habit of excess animal protein. The old people in the blue zones are essentially vegetarian and only sometimes eat meat [1].

10.2   Methionine Hypothesis

10.2.1   Toxic Effects of Methionine from Excess Animal Proteins

In the Europe and the USA, many people have died of myocardial infarction over the past century and the reason may have been due to excessive intake of animal proteins. Many researchers have studied which amino acids are responsible for myocardial infarction in the past 70 years and found that the reason must be methionine, which is at a high concentration in animal proteins. Biochemically, excess methionine intake leads to an increase in homocysteine, a toxic amino acid, which is suspected to be the real killer of animal proteins. More homocysteine is produced in the liver with a greater intake of animal proteins (methionine). In particular, homocysteine is now proposed to be the real cause of myocardial infarction [2]. Although overconsumption of animal protein to decrease carbohydrates may cause good results in the short term, this will increase homocysteine and increase the risk of myocardial infarction in the long term [3]. In addition, although this is just an animal experiment, the restriction of methionine (effectively corresponding to the restriction of meat) extended to worms, flies, and even mice. To these backgrounds, the hypothesis that the methionine-homocysteine pathway may play a central role in aging is called the "methionine hypothesis" [3].

In 2018, methionine restriction in mice could even extend lifespan (Fig. 10.1). Nevertheless, it could not be concluded that methionine restriction extends the lifespan of *Homo sapiens*. However, too much intake of methionine (animal proteins) confers possible disadvantages and I do not agree with the opinion on excess animal protein consumption [3].

In contrast, animal proteins cannot be too much. A complete vegetarian may cause big disadvantages. This is due to the decrease in tryptophan, one of the essential amino acids. Tryptophan is a precursor of serotonin, the "happiness hormone." A serotonin deficiency is supposed to be the main cause of depression. Strict vegetarians have a significantly high percentage of depression.

In summary, animal proteins are essential, but the problem is the amount that should be consumed: in Japan, 50 g (woman) and 60 g (man) protein are

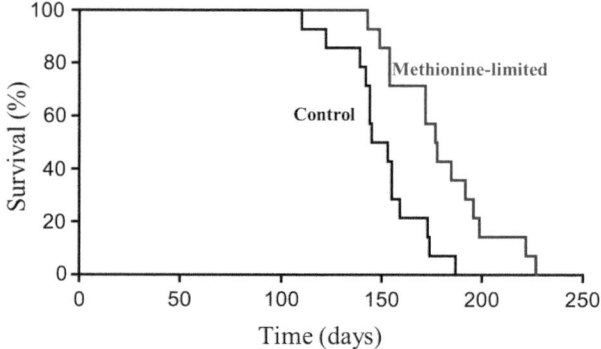

Fig. 10.1 Extended lifespan of mice by methionine-limited diet suggesting that methionine of animal protein may limit longevity [3], suggesting that too much animal protein (too much meat) is harmful to human longevity. Too much intake of meat may induce cardiac failure because too much methionine may induce the accumulation of homocysteine in the systemic circulation [4, 5]

recommended by the Japanese Ministry of Health, Labor, and Welfare; unfortunately, I have no quantitative data on how much animal protein should be taken. Maybe, the preferential amount of meat is up to 100 g/day. We can consider the following two standards on how much animal protein to take.

1. Not too much methionine [4, 5]
2. Not too little tryptophan [6, 7]

The wisdom of eating habits to satisfy both standards are in two books, "*The Blue Zones*" and "*Japanese Villages of Longevity and Short Life*" written by Dr. Dan Buettner (USA) [1] and Prof. Shoji Kondo (Japan) [8], respectively. They do not have extremely biased opinions (extreme carbohydrate restrictionists or complete vegetarians).

We should not take too much methionine by increasing animal proteins to increase ketone bodies (extreme carbohydrate restrictionists) or decrease tryptophan by decreasing animal proteins (complete vegetarians). We can be majorly disadvantaged by these extreme opinions.

10.2.2 Serotonin Hypothesis and the Gut-Brain Axis

The serotonin hypothesis is the viewpoint that puts serotonin at the center of "the brain feels happy." The serotonin hypothesis is of high value, especially to humans, who have large brains. Happiness is not measured by the use of worms and flies. In human life, it is very important to examine the effects of serotonin and how serotonin is involved in the feeling of happiness [9]. Feeling happy in daily life is the most important element for human longevity. This is why studies by use of worms, flies, and mice cannot fully explain human longevity.

In happiness, serotonin works well in the brain and drives emotions upwards. In contrast, serotonin does not function at all in the brain and drives emotion downwards into depression. LysergSäureDiethylamid (LSD), which is designated as a psychotropic drug, binds to and activates the serotonin receptor, and has a strong hallucinogenic effect. It even causes fatal accidents due to confusion. This shows how serotonin has a definitive role in the brain. Serotonin has a big impact on emotions, but also on pain, feelings, facial expressions, and body conditions. In the recent scientific results, serotonin has a close relationship with eating habits, termed the "gut-brain axis" (Fig. 10.2).

Depression is at least partly due to the destruction of the gut-brain axis. If the gut-brain axis is disrupted, the chances of becoming "depressed" increase endlessly. This is the reason why people who always eat junk food get depressed easily. In contrast, improvements in eating habits may lead to recovery from depression.

Recent studies show that the activities of the gut (enterobacteria) and the brain (mental activities) are closely related. People with good gut microbiota have many chances of living a happy life and people with poor gut microbiota have a low probability of living a happy life. In contrast, people who think that they are happy may have a good microbiota and people who think they are unhappy may have a poor microbiota [9].

The serotonin system may be one of the possible linkers between gut microbiota and mental activities because a precursor of serotonin is synthesized in the gut and serotonin is the key to feeling happy. Serotonin synthesis is initiated by tryptophan, which is produced in the gut by the hydrolysis of proteases. Tryptophan is converted to 5-hydroxytryptophan by a specific enzyme in gut tissues and is delivered to the brain through the blood-brain barrier. Please note that 5-hydroxytryptophan can

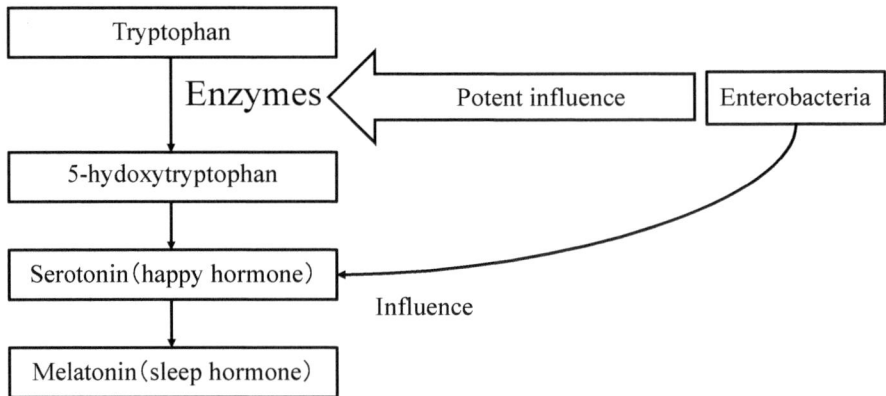

Fig. 10.2 Gut-brain axis mediated by tryptophan-serotonin pathway, suggesting that a good microbiome is essential for feeling happiness in daily life [9]. Since 5-hydroxytryptophan, a precursor of serotonin, is synthesized in the gut, the good environment of the gut is essential to maintain serotonin levels in the brain. In turn, serotonin secreted from the nerve end of the gut regulates the movement of intestinal smooth muscles and the secretion of digestive enzymes. These interactions are termed as the "gut-brain axis." Therefore, a good gut environment is an essential factor to feel happy in human daily life

target the brain much more effectively than tryptophan. Consequently, serotonin synthesis depends on the gut because over 90% of 5-hydroxytryptophan is synthesized in the gut (Fig. 10.2). The enzyme for the hydroxylation of tryptophan is in the gut endothelium and finally the synthesis of serotonin (a key molecule to feeling happy in daily life) is strongly influenced by the gut microenvironment. According to mice experiments, the transplantation of good microbiota can greatly improve depression [9].

5-Hydroxytryptophan synthesized in the gut can be transferred to the brain by blood flow and can pass freely through the blood-brain barrier. Thus, this can target neurons, be converted to serotonin, and finally have various functions in the brain. Serotonin can make the emotion upward, and the brain can feel happy. Furthermore, serotonin can be converted to melatonin in neurons and function as a sleep-inducing substance in the brain. Serotonin is high in the daytime and melatonin is high at night. Through these hormones, humans may generate a sound circadian rhythm; humans are active in the daytime and relaxed at night. It may be argued that a good circadian rhythm formed by serotonin (happiness hormone) and melatonin (sleep hormone) is the basis of a good human life. I am looking forward to seeing how the serotonin hypothesis will progress in the future [9].

10.3 Healthy Longevity

10.3.1 Serotonin, Ketone Bodies, and Microbiota

Dr. Yuji Naito, Prof. of Kyoto Prefectural University of Medicine reported a very interesting result. The old people with healthy longevity in Kyotango have significantly more butyrate-producing enterobacteria than those in Kyoto city [10]. The gut environment of abundant butyrate-producing enterobacteria has short-chain fatty acids, such as acetic acid, propionic acid, and butyric acid, and can activate regulatory T cells in the Peyer's patch in the gut, resulting in the suppression of chronic inflammation in the whole body. In addition, such an environment enhances the hydroxylation of tryptophan, possibly to increase serotonin in the brain. It may be reasonable that residents of the Longevity Village are willing to do farm work while having fun, suggesting that good gut microbiota are closely related to their mental activities. The residents in the longevity may have the ideal gut-brain axis [11].

Neurons expressing serotonin receptors accumulate in the frontal lobe and deeply affect human emotion. These neurons are outstandingly large compared with other neurons and accept vast input from them. These neurons are in similar situations to pyramidal neurons. They require a large amount of energy to do their jobs. When energy is not in abundance, they do not perform their task. Neurons expressing serotonin receptors may critically require ketone body shuttle. Ketone bodies can be synthesized in the surrounding astrocytes, as well as by hepatocytes in the liver.

Because the liver has a vast capacity for ketone body production, it may be preferential for activating the ketone body shuttle in the brain. In conclusion, for the brain to work well, the pyramidal neurons activated by serotonin should be continuously supplied with energy.

If the frontal lobe neurons can be operated at full power, the function of serotonin will be maximized. The emotion will always be active. Good sleep will be possible due to an increase in melatonin.

So, let us consider what we should do by summarizing the topics. Meat is essential, but not too much, maybe up to 100 g/day. Vegetables are also essential, but a strict vegetarian diet is not recommended. Serotonin functions should be maximized by keeping the gut environment healthy. Since neurons with serotonin receptors as well as pyramidal neurons with glutamate receptors are much bigger than granular neurons, they are highly sensitive to a shortage of energy substrate. Therefore, an increase in ketone bodies supplied by hepatic cells or astrocytes may be effective in preserving their activities. The most important point is that we must consider our eating habits in total balance.

10.3.2   Moderate Eating Habits Keep the Brain Active

In addition, the maximization of serotonin's power is not recommended to realize healthy longevity. What we need in daily life is to keep serotonin continuously active. These states of the brain may contribute to the mentality of the old people in villages of longevity. They are enjoying their daily life and farm work. According to their daily eating habits, the concentrations of ketone bodies may be 0.2–0.5 mM (small ketogenic). The range of concentrations of ketone bodies is far from those required for fully powered serotonin but is sufficient for keeping the brain calm and constitutively active.

In addition, it is within the range of the ketone body concentrations (0.2–0.5 mM) that Mary T. Newport described as improving the cognitive function of her husband in "*Alzheimer's Disease: What If There Was a Cure? The Story of Ketones*." For the improvement of cognitive functions, this range is at least sufficient [12]. For example, the range of ketone body concentrations can significantly activate the HCAR2.

We can attain healthy longevity by improving our eating habits to yield a very slight (0.2–0.5 mM) increase in ketone bodies. The big advantage of this eating habit is obtained by continuing it for our life. Through this system, ketone bodies account for less than 10% of all energy demand and anybody can pursue this hybrid system of glucose and ketone bodies.

According to Prof. Shoji Kondo (1893–1977), the pioneer of studies on villages of longevity, the basis of healthy longevity is coarse food and hard labor. Healthy old people over 100 years old also perform physical labor. The old residents of the Longevity Village eat coarse food and do not have refined sugar or too much meat and their main food is vegetables [8].

These eating habits do increase ketone bodies, but the increase should be very small. In conclusion, the solution to healthy longevity is not to increase ketone bodies (stringent carbohydrate restriction) or to eliminate ketone bodies. The solution is a moderate life, which is the situation in which the brain is kept active. A possible solution for ketone body concentrations (0.2–0.5 mM) is one that anybody can maintain every day. You can continue without overdoing it; you cannot eat only meat and or be strictly vegetarian. You should practice a diet that incorporates what you like [8].

10.4   A Man Is as Old as His Arteries

10.4.1   The Common Features of Descendants of Centenarians

Descendants of centenarians were compared with the control group to determine the secrets of healthy longevity. This study aimed to identify the traits of longevity that are passed to their offspring through epidemiological tests. There were no significant differences in cerebral infarction, arrhythmia, hypercholesterolemia, diabetes, dementia, cancer, osteoporosis, and osteoarthritis (Table 10.1). There was a significant difference in hypertension and arrhythmia. What can we learn from this study? Notably, hypertension and arrhythmia are due to the loss of flexibility of blood vessels. The key to centenarians' longevity may be retaining the flexibility of blood vessels. On the contrary, the loss of flexibility of blood vessels leads to aging.

It should be noted that centenarians' descendants are compared with control groups in the study. This is because only the "flexibility of blood vessels" is inherited by the offspring among all the traits common to centenarians. People who lose the flexibility of blood vessels have high blood pressure due to the strong beating of the heart.

Table 10.1 Offspring of centenarians' longevity have a lower prevalence of hypertension and arrhythmia [13, 14]

Disease	Offspring of control group (%)	Offspring of centenarians (%)
Myocardial infarction	11	6
Cerebral infarction	6	2
High blood pressure	63	37 (Significant difference)
Arrhythmia	23	10 (Significant difference)
High cholesterol plasma	47	40
Diabetes mellitus	10	10
Dementia	1.3	0.5
Gun	20	11
Osteoporosis	26	19
Osteoarthritis	56	44

I recall the quote by Dr. William Osler "A man is as old as his arteries"—suggesting that humans age as blood vessels age. Please consider that over half of deaths are caused by failure of blood vessels, including stroke, heart failure, and renal failure.

In conclusion, keeping blood vessels flexible is the key to living for more than 100 years. Let us discover the secret method in the next section.

10.5   Longevity Research in Japan

10.5.1   Conclusions from the Conversation with Healthy Old

The great pioneering work of Dr. Shoji Kondo, studying the villages of longevity and short life, should be considered [8]. I obtained this book 10 years ago and have it to hand. I argue that this is the best study of antiaging. A series of research was performed 70–100 years ago, although the book was edited in 1972, and, back then, he was the first person who had noticed that too much carbohydrate induces short life. His conclusion is highly simple and easily understandable to everybody. Surprisingly, he reached the same conclusions with the "insulin hypothesis" [13, 14], "serotonin hypothesis" [6, 7], and "methionine hypothesis" [3, 4] as the current aging hypothesis by his original fieldwork. He visited the villages of longevity and villages of short life all over Japan before World War II (1920s–1940s) and searched for the essence of longevity and short life of the Japanese people. He directly questioned elderly individuals on their eating habits and lifestyle and extracted the physiological essence of aging. Back then, no other researchers studying physiology and public health cared for aging research. However, his pioneering work has contributed to the current aging research all over the world. The formal title of the book is *"Green Vegetable, Seaweed and Soy-determined Villages of Longevity and Short Life in Japan."* This is the name given by the editor of this book, Dr. Hiromichi Ogiwara. I have read this book multiple times and found that he emphasized the following three points [8]:

1. Do not eat too much carbohydrate (by the current words "insulin hypothesis")
2. Do not eat too much meat (by the current words "methionine hypothesis")
3. Eat vegetables, seaweed, and soybeans little by little every day

Because items 1 and 2 were not approved by most researchers in the 1960s (when the book was first published), I suspect Dr. Ogiwara chose item 3 as the book title. In the 1960s, a book on carbohydrate restriction would not have been accepted in Japanese society.

Dr. Kondo used a very simple and direct method. He asked the elderly people of the villages a series of questions on eating habits and lifestyle and extracted general principles from their answers. This method must have demanded much endurance and imagination. Back then (1920s–1960s), aging theory was so primitive that no researcher was able to explain his conclusion with the advanced aging theory of the time. Here, I explain his conclusion by the current aging theory [8].

"My specialty is hygienic, not nutrition. I have studied hygienics for a long time and found the problem of longevity in the Japanese. This is 1927. During the days of this study, I am determined to continue this work and I have been devoting myself to this study."

"Anybody did not find this problem when I found this problem (healthy longevity) in 1927." From the researcher's perspective, "I will explore what anybody does not find" concerns the originality of the use of specific language. This is a clear indication of Prof. Kondo's outstanding nature. I can only respect his courage to devote himself to a very new area of study: "Do not conclude anything on your desk without experiments or practice or observation. I have visited villages of longevity and villages of short life all over Japan with the feeling that I don't conclude without collecting examples in practice. Finally, I reached the first conclusion" [8].

Whether the village of longevity or the village of short life is determined by eating habits that have continued every day for a long time. This is an important warning to young scientists. They cannot determine anything by reading papers alone. The importance is a conclusion drawn through experiments and practices. The conclusion evidenced by experiments may contribute to human life. In addition, Dr. Kondo concluded that eating habit is the main priority for antiaging. At present, this has become established, but it was Dr. Kondo who reached this conclusion through a vast amount of fieldwork [8].

"Anyone who has eaten only rice since he was young will die young. For example, the areas of rice production in the North–East Japan are commonly seen." "Especially, people in the Akita prefecture, which have many villages of short life, are eating only rice. They are eating much white rice with salty pickled radish in miso and eggplant pickled in miso." "Looking up the family register of the government office, most people collapse in stroke in their 60s. Since they are rapidly aging, they give up farm working in their 50s." This is a pattern of death by stroke due to the habit of eating too much carbohydrate. According to the insulin hypothesis, there must be sequential events by prolonged and repeated insulin spikes during the habit of eating too much carbohydrate. Excess carbohydrate-glucose spikes-insulin spikes-blood vessel aging [8].

"I visited the villages where the residents have to work hard to examine whether lifespan is shortened by hard labor. I found it interesting that the results are contrary to this hypothesis. The villages of hard labor often turned out to be those of longevity." "That is the island of longevity where the old people are healthy and working hard. The 90-year-old men are working on the farm and the women are weaving a thin obi, a specialty of Okinawa" [8].

These results have a very important suggestion. Many studies on longevity and blue zones have now reached the same conclusion. Surprisingly, Dr. Kondo had noticed it 100 years ago. Most of the residents in villages of longevity eat coarse food and perform hard labor. The old people are commonly having fun and working on their daily farming work that usually requires hard labor. The current brain research suggests the same conclusion, namely that daily life with adequate levels of serotonin is a condition for longevity. In addition, the following statements have more interesting suggestions [8].

10.5.2   The Residents of Fishing Villages Eat Seaweed Every Day

"After I surveyed many fishing villages, I found that residents in some of them are eating only fish. Of course, they buy rice to eat. Residents in fishing villages without a farm have this eating habit. They eat only fish and little vegetables. People with this eating habit do not fail to die young." "I was looking into the death of dying young due to the eating habit and found that most of them died of angina, myocardial infarction, or heart attack" [8]. The cause of death can be explained by the methionine hypothesis. Methionine is abundant in animal proteins such as fish, chicken, cows, and pigs, and is metabolized in the liver to produce homocysteine, a toxic amino acid that is supposed to be the cause of heart failure. Too much animal protein or meat should be avoided to prevent heart failure [2, 3].

"I found that people who ate seaweed daily hardly died of stroke, even though the residents of the surrounding rice-farming villages died young of stroke, for example, Toga village on the Oga peninsula in the Akita prefecture. There is a clear distinction between neighbor villages. These effects must be due to dietary fiber, such as alginic acid and fucoidan, in seaweed, which has favorable effects on gut microbiota. In the current terminology, these are known as prebiotics" [8].

Finally, Dr. Kondo mentioned how elderly people eat their food. Anybody may agree with this comment. "This is justified by the survey of the villages all over Japan. One shot of eating seaweed never has a healthy effect, one must have an eating habit. Every day one has seaweed little by little." Eating small amounts of vegetables and seaweed relatively often may be a good habit for the brain [8].

10.6   Removing Bad Habits

10.6.1   The Story of My Hometown

Why do humans age? In response to this question, there is sometimes a simple answer. For example, a heavy smoker or a person addicted to alcohol. What the person must do is to stop bad habits. He should give up smoking or drinking. In this case, he does not have to consider why he is aging. I grew up in the South Iwate prefecture, where the rural landscape of rice production is abundant. It fits perfectly with the conditions of a "short-living village" defined by Dr. Kondo [8]. The residents ate too much white rice every day and made much rice cake at every event to eat for celebration. Hence, they consumed excess carbohydrates. Most of the same generation as my father were smoking every day and died of lung or esophageal cancer. In addition, many of the same generation as my grandfather drank a liter of Japanese sake each day. However, they suffered from high blood pressure and were even proud of their high blood pressure (over 200 mmHg). They died of a stroke. In my hometown, so many persons died of stroke that there was a dialect "he was hit." When I grew up, I became aware that "he [someone] was hit" in south-northeast Japan, whereas "he hit it" in north-northeast Japan. When my parents wore mourning dresses and went out, they often said "he was hit to death," which is a dialect version of "he died of a stroke." Because the town is a rice-producing area, the residents ate too much steamed white rice. As Dr. Kondo stated, "A rice-producing area is a village of short life without an exception because many died of stroke" [8].

My father died of small cell carcinoma, a type of lung cancer, one of the most malignant cancers, 12 years ago. He smoked two boxes of cigarettes a day. At some point, he could not stop coughing for more than a week and suspected that he would have pneumonia. He drove himself to a general hospital in the area about 15 km away. The next, day, I was called by a young doctor of respiratory medicine. He said, "He has small cell carcinoma, Stage IV. The patients have a very short life expectancy. In addition, he has almost no platelets. I had a similar patient 3 years ago and he died 7 days later." My father died on a cold morning, 10 days later. When a patient is about to die, I learned that the doctor examines the color of their urine. When the kidney stops functioning, urine becomes dark wheat-colored (raw urine). The patient's heart simply stopped after a few hours of beating and the end of life was declared. The death several hours later is confirmed when kidney function stops. The kidney may sometimes determine how long a person lives.

The patients hospitalized for respiratory reasons were all middle-aged or old men and all were heavy smokers in the rural area of Japan. Their disease was chronic obstructive pulmonary disease (COPD) or lung cancer. Many patients were left the hospitals by the hearse as a corpse. Even if they were lucky enough to be discharged, they would leave the hospital in a wheelchair with an oxygen cylinder. As I write this section, I find myself wishing that heavy smokers would give up smoking after reading my statements here.

10.6.2 Aging Research as Chaos

Although I have digressed, there are three reasons why my hometown was a village of short life.

1. Heavy smoking
2. Excess drinking
3. Excess carbohydrate

Giving up these three habits may lead to longevity. These may be the basic conditions for longevity. I used to agree with this opinion, but now I consider it differently. These days, most people do not smoke, drink a little, and save carbohydrates. There is a problem that even these people are aging.

Even now, many theories of aging are proposed. It is necessary to ask the basic question of why humans are aging. If there were 100 researchers, there would be 100 answers to this question. The current situation is that we do not know what will happen tomorrow, even with the most influential hypothesis of aging supported now by many researchers. The study of aging is "a world of chaos."

References

1. Buettner D. The blue zones. National Geographic; 2012.
2. Martí-Carvajal AJ, Solà I, Lathyris D, Dayer M. Homocysteine-lowering interventions for preventing cardiovascular events. Cochrane Database Syst Rev. 2017;8(8):CD006612.
3. Bárcena C, Quirós PM, Durand S, Mayoral P, Rodríguez F, Caravia XM, Mariño G, Garabaya C, Fernández-García MT, Kroemer G, Freije JMP, López-Otín C. Methionine restriction extends lifespan in progeroid mice and alters lipid and bile acid metabolism. Cell Rep. 2018;24(9):2392–403.
4. Wanders D, Hobson K, Ji X. Methionine restriction and cancer biology. Nutrients. 2020;12(3):684.
5. Gao X, Sanderson SM, Dai Z, Reid MA, Cooper DE, Lu M, Richie JP Jr, Ciccarella A, Calcagnotto A, Mikhael PG, Mentch SJ, Liu J, Ables G, Kirsch DG, Hsu DS, Nichenametla SN, Locasale JW. Dietary methionine influences therapy in mouse cancer models and alters human metabolism. Nature. 2019;572(7769):397–401.
6. Fidalgo S, Ivanov DK, Wood SH. Serotonin: from top to bottom. Biogerontology. 2013;14(1):21–45.
7. Dell'Osso L, Carmassi C, Mucci F, Marazziti D. Depression, serotonin and tryptophan. Curr Pharm Des. 2016;22(8):949–54.
8. Kondoh S. Long- and short-lived villages in Japan. Sanroad Publishing; 1991.
9. O'Mahony SM, Clarke G, Borre YE, Dinan TG, Cryan JF. Serotonin, tryptophan metabolism, and the brain-gut-microbiome axis. Behav Brain Res. 2015;277:32–48.
10. Naito Y, Takagi T, Inoue R, Kashiwagi S, Mizushima K, Tsuchiya S, Itoh Y, Okuda K, Tsujimoto Y, Adachi A, Maruyama N, Oda Y, Matoba S. Gut microbiota differences in elderly subjects between rural city Kyotango and urban city Kyoto: an age-gender-matched study. J Clin Biochem Nutr. 2019;65(2):125–31.
11. Agirman G, Yu KB, Hsiao EY. Signaling inflammation across the gut-brain axis. Science. 2021;374(6571):1087–92.
12. Newport MT. Alzheimer's disease: what if there was a cure?: The story of ketones. Basic Health Publications; 2011.
13. Brugts MP, van den Beld AW, Hofland LJ, van der Wansem K, van Koetsveld PM, Frystyk J, Lamberts SW, Janssen JA. Low circulating insulin-like growth factor I bioactivity in elderly men is associated with increased mortality. J Clin Endocrinol Metab. 2008;93(7):2515–22.
14. Brugts MP, van Duijn CM, Hofland LJ, Witteman JC, Lamberts SW, Janssen JA. Igf-I bioactivity in an elderly population: relation to insulin sensitivity, insulin levels, and the metabolic syndrome. Diabetes. 2010;59(2):505–8.

Chapter 11
Brain Energy System Against Neuronal Diseases

Abstract Have you ever heard of the disease of Epilepsy? The patients sometimes have seizures with convulsions and usually possible management of symptoms by drugs. However, some patients cannot be managed by drugs. Epilepsy is the energy problem of the brain. Further, it is closely related to the problem of energy substrate ATP. When using a computer, you have certainly experienced a freeze. The computer stops responding to whatever you should try. A computer has a point of informational junction. When the point is down, the computer will be frozen. As the exchange of information is performed through electric signals in a computer, communication between neurons is made possible by them. In addition, the brain has a point of information junction as a computer has. The point is pyramidal neurons, which are larger than other cells, have more dendrites, receive more input, and send out more signals. Many pyramidal neurons are excitatory in response to glutamate. These types of neurons have safety devices that suppress their electric signals. Whether or not the safety devices can be activated is the key to protecting this huge neuron (excitatory neuron). Epilepsy patients cannot activate this safety device. If the drug is not effective, there is only one way to control a special diet called the ketogenic diet. Here I talk about why ketone body can improve symptoms of epilepsy. By understanding this, I will let you know that ketone bodies may be a final hope for the brain.

11.1 Epilepsy

11.1.1 Rescue of Refractory Epilepsy by Ketogenic Diet

The brain is the organ in which the harmful effects of an excess carbohydrate diet are most likely to appear. The diet may cause brain energy problems [1, 2]. This is especially true of epilepsy, which has a certain prevalence in children. As seizures usually occur every several months, many people can live an ordinary life with anti-epileptic drugs. However, refractory epilepsy is a different story. The patients have seizures several times per day and no drugs are effective. Although there are no effective drugs, the ketogenic diet is known to have clear therapeutic effects on refractory epilepsy. The ketogenic diet is a special food that contains 80% of total energy as fat and several grams of carbohydrates. When adhering continuously to this diet, ketone body concentration is increased to over 5 mM, approximately 50 times higher than normal, a concentration that can completely prevent seizures. This is why ketogenic diets have now recaptured the attention of many people in the USA and Japan [1–3].

It is said that lifestyle-related diseases, including epilepsy and diabetes, may be caused by carbohydrate-centered eating habits, which developed when *Homo sapiens* started farming 10,000 years ago. Ketone bodies are possible tools to overcome the negative effects of this carbohydrate-centered eating habit. Overall, I am certain that ketone bodies are the last lines of defense for the brain.

11.1.2 Shutdown of Seizures by 5 mM Ketone Bodies

Usually, when blood glucose levels are stable, brain energy is completely dependent on lactate and glucose supplied from astrocytes. However, the standard Japanese diet does not increase ketone bodies at all. This is why ketone bodies are a sub-engine providing for neurons in an emergency.

When ketone body concentration is increased, the brain is ready to use ketone bodies as the most effective energy substrate. According to old papers on fasting physiology [3], an increase in the blood concentration of ketone bodies increases its contribution to the brain's energy metabolism (Table 11.1). In a normal diet, the ketone body concentration is 0.1 mM and it does not contribute to the brain energy. When increased to 1.0 mM, the contribution is around 30%, and when increased to 5.0 mM, the contribution is 70%. In the case when ketone bodies account for 70% of the brain's energy, any epileptic seizures are completely prevented.

Since epileptic patients account for 1% of the population, there are approximately 1 million patients in Japan. Among them, 30% of patients are refractory (i.e., no drug is effective against seizures), meaning that 300,000 people experience repeated seizures. These patients should increase ketone body concentration to suppress epilepsy. They should aim for ketone body concentrations of approximately 5 mM [4, 5].

Table 11.1 Contribution to brain energy consumption at various ketone body concentrations [3]

Ketone body concentrations in the blood	Brain energy by glucose	Brain energy by ketone bodies
0.1 mM (normal diet)	100%	0%
0.2–0.5 mM (calorie restriction)	>90%	<10%
1.0 mM (carbohydrate restriction)	70%	30%
2.5 mM (fasting for several days)	50%	50%
5.0 mM (continuous ketogenic diet)	30%	70%

11.2 Suppression of Excessive Excitatory Potentials

I report here the relationship between ketone body concentrations and seizure frequency, based on a patient's experience (Table 11.2). This is not a static analysis but an experience of an epilepsy patient. Thus, this cannot be generalized but is presented here for the reader's reference. At present, several devices are available to monitor ketone bodies for home use. Anybody can measure their blood concentration of ketone bodies and glucose at home. The relations between ketone body concentrations and inhibition of seizure are almost the same as those obtained in basic science by use of mouse or rat brain slices [4].

According to Table 11.2, when the ketone body concentration was increased to 2.0 mM, the frequency of seizures was suppressed. From Table 11.1, it is suggested that the contribution of ketone bodies to the brain energy metabolism is around 50%. Perhaps when ketone bodies contribute about half of the brain's energy, seizures are suppressed, even in the case of refractory epilepsy.

Please see Fig. 11.1; a slice of mouse neocortex has been cultured, inducing excessive excitatory action potentials. In vitro, cultivation easily induces pathological conditions in which the brain has several hundreds of action potentials per second.

However, if these action potentials are induced in real life, this would be a major disaster for the brain. In particular, this would cause a serious situation in pyramidal neurons that usually have a high energy demand. This is an epileptic seizure. First, the intracellular ion composition changes. The ion composition of neurons will be changed. The most serious problem is the accumulation of Na^+. It is a Na^+ pump that extrudes this ion from neurons. Neurons fully activate the Na^+ pump and try to remove excess Na^+.

In a healthy brain, the intracellular energy substrate ATP decreases and activates the K^+ ATP channel. This K^+ ATP channel is such an effective safety device that pyramidal neurons can suppress the excessive action potentials. The brain remains safe and is rescued from trouble by the assistance of the K^+ ATP channel. However, patients with epilepsy cannot activate the K^+ ATP channel. The excessive action potentials spread all over the brain. This is a scientific explanation for an epileptic seizure. Please confirm that a single line corresponds to a single action potential. In the absence of 3-hydroxybutyrate, so many action potentials are induced you can see a black column.

Many researchers report that a 3-hydroxybutyrate (3HB) concentration of 2–3 mM can inhibit excessive action potential by the use of brain slices (Fig. 10.2). Before the addition of 3HB, the black columns indicate the repeated action potentials. In the presence of 3HB (2.5 mM), action potentials are completely

Table 11.2 Frequency of seizure and ketone body concentration

Ketone body concentrations	Frequency of seizure
<0.5 mM	Negligible effect
Around 1.0 mM	Considerable decrease
Around 2.0 mM	Significant decrease
Around 5.0 mM	No seizure

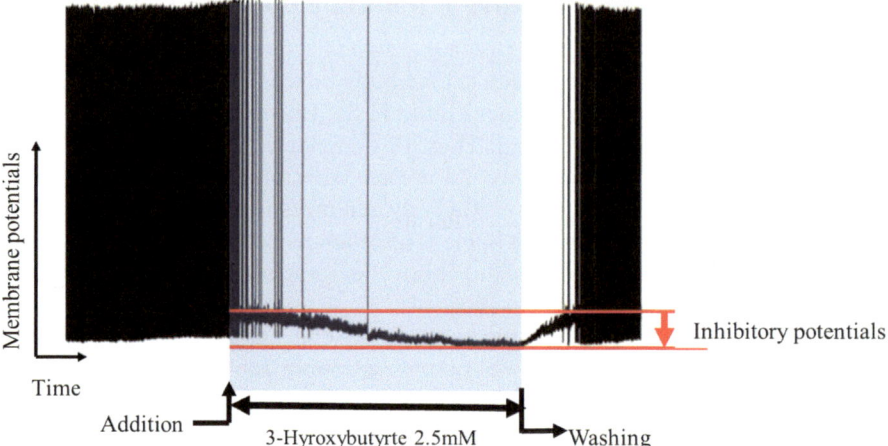

Fig. 11.1 3-Hydroxybutyrate (3HB: the major compound of ketone bodies) inhibits excessive action potentials [4]. Note that 3HB has two actions here. One is the inhibition of action potentials, and the other is a decrease of resting potentials. This is the critical feature of K⁺ATP channel activation. Even now, only a ketogenic diet is a possible method to maintain ketone body concentrations over 2.5 mM for patients of refractory epilepsy to live a normal life

inhibited, and the baseline of potential (named as the resting potential) is considerably decreased. Usually, the resting potential is −80 mV, but 3HB effectively decreases the resting potential by approximately 10 mV. This leads to complete inhibition of action potentials [4, 5].

The decrease in resting potential is caused by the activation of the K^+ ATP channel. Thus, please consider how the K^+ ATP channel is effective against seizure. As far as I know, there are no activators of this K^+ ATP channel other than 3HB. This is the reason behind this chapter. I would like the readers to realize the importance of ketone bodies as the last defense line against refractory epilepsy. When 3HB was washed out, repeated action potentials returned [4, 5]. 3HB concentration increased to 2.5 mM can activate the K^+ ATP channel, decrease the resting potential, and decrease the frequency of action potentials. This is why ketone bodies rescue pyramidal neurons from overwork [4, 5].

11.3 K^+ ATP Channel

11.3.1   Two Systems for Inducing Inhibitory Potentials

There are major two types of neurons: excitatory neurons (the neurotransmitter is glutamate) and inhibitory neurons (the neurotransmitter is GABA). Since the human brain is large, there is a high risk of excessive excitatory potentials starting from excitatory neurons. To keep the brain calm at any time, humans must protect their safety devices against excessive excitatory potential. For this purpose, the brain has inhibitory neurons. This type of neuron exists to suppress action potentials. These systems reduce resting potentials by about 10 mV. The following two systems occur for inhibition:

1. Inhibitory inputs are mediated by GABA-inhibitory neurons [6]
2. Safety devices in excitatory neurons mediated by the K⁺ATP channel [4, 5]

GABA is a well-known neurotransmitter that induces inhibition of action potentials. In addition, GABA has become a popular food ingredient in sweets, including chocolate. It is said that GABA can stabilize emotions by these inhibitory actions on excitatory action potentials. GABA binds to a specific receptor on neurons, induces inhibitory potential, and inhibits action potentials. If GABA is injected into the mouse brain, action potentials are reduced and maybe emotions are stabilized. A common tranquilizer used in neurological fields is an activator of the GABA receptor. In humans, GABA cannot directly be injected into the brain; instead, drugs are used that activate the GABA receptor. Diazepam and clonazepam are well-known tranquilizers are GABA receptor activators, but they have limited effects on refractory epilepsy, or may not be effective at all.

Excitatory neurons themselves have a safety device to stabilize membranes (Fig. 11.2). This is the K⁺ ATP channel. In addition, 3-hydroxybutyrate is the most potent activator of the K⁺ ATP channel. When the K⁺ ATP channel is activated, the membrane potential is decreased so that neurons themselves inhibit action potentials without inhibitory input. *Homo sapiens*, with a large brain, face a high risk of danger from evoking excess action potentials from any part of the brain. Thanks to the K⁺ ATP channel, the brain can inhibit the spread of action potentials. For

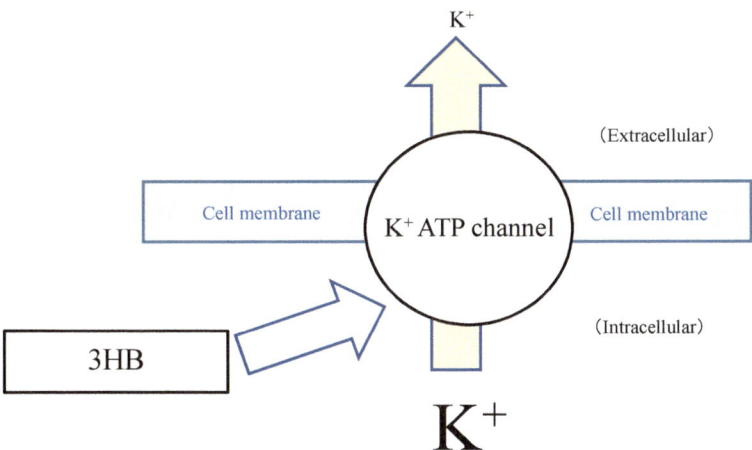

Fig. 11.2 The K⁺ ATP channel is a stabilizer of pyramidal neurons that rescues pyramidal neurons from overwork. 3HB activates the K⁺ ATP channel possibly through direct binding to the channel protein and inhibits evoking a series of action potentials. Thus, 3HB can suppress evoking excessive action potentials

large-brained humans, the K^+ ATP channel is an important safety device. The key to effective use of this safety device is pyramidal neurons that play a role in the informational hub. In addition to this inherent safety device in excitatory neurons, neurons can receive inhibitory input from GABA neurons. The K^+ ATP channel is a much more effective stabilizer of membrane pyramidal neurons than GABA neurons [4, 5].

11.3.2   Inactivation of K^+ ATP Channels in Patients with Refractory Epilepsy

Thus, to combat seizure, the activation of the K^+ ATP channel is a much more effective measure than the activation of GABA neurons. To keep the brain calm, it is more important for us to focus on the K^+ ATP channel in pyramidal neurons. The following two events are closely related to the activation of the K^+ ATP channel.

1. Decreased ATP
2. Increased ketone body concentration

The K^+ ATP channel (a type of K^+ channel) is such an effective safety device that excessive action potential will not be sustained if K^+ ATP is normally activated. The K^+ ATP channel is highly unique in that this is activated when the energy substrate ATP is decreased. Since K^+ ATP is usually potently suppressed by energy substrate ATP, this can be understood in terms of neuronal energy metabolism. The energy substrate ATP binds to K^+ ATP and inhibits K^+ ATP. When there is a sufficient amount of ATP, the K^+ ATP channel is completely inhibited because K^+ ATP does not usually need to work. In contrast, when the amount of energy substrate ATP is decreased (i.e., in an emergency), the K^+ ATP channel is activated and completely inhibits action potentials. This is how the K^+ ATP channel works for the brain as an effective safety device. By activation of the K^+ ATP channel, the membrane potential is sharply decreased due to efflux of K^+ ions to the extracellular space. Finally, the neuronal membrane will stabilize as action potentials are less likely to occur.

Just after the action potentials repeatedly occur, a large amount of energy substrate ATP is consumed. The action potential itself does not demand energy consumption. However, Na^+ influxes into neurons should be immediately extruded by the Na^+ pump. Neurons must consume large amounts of energy substrate ATP to keep the Na^+ pump working well.

The K^+ ATP channel is a safety device that rescues neurons from this burden. The K^+ ATP channel demonstrates the excellent feature of taking the initiative and suppressing the action potential. The K^+ ATP channel becomes active in response to a decrease in energy substrate ATP just below the membrane, decreases membrane potential, and inhibits action potentials. This saves energy consumption and suppresses the spreading of excess action potentials in the brain. Thus, seizures are completely inhibited in the healthy brain. Through this system, neurons do not fall into overwork and always perform optimally well [1, 2].

However, patients with epilepsy do not have a functioning K⁺ ATP channel. In some cases, the excessive action potentials can spread easily over their brain. This is because the K⁺ ATP channel cannot be activated, even when the energy substrate ATP decreases. When a neuron causes repeated action potentials, neighboring neurons cannot activate safety devices and, finally, repeated action potentials spread out all over the brain. This is an epileptic seizure [1, 2].

11.4   Activation of the K⁺ ATP Channel

11.4.1   The Modes of Action to Stabilize Membrane Potentials by Ketone Bodies

Why do ketone bodies activate the K⁺ ATP channel to suppress excess action potentials? It is not a clear situation. The three effects of ketone bodies should be studied in the future.

1. Direct binding to the K⁺ ATP channel
2. Activation of the HCAR2 receptor on the cell membrane
3. Activation of mitochondrial metabolism

The first possibility is that ketone bodies directly bind to the K⁺ ATP channel to activate the channel. In addition, ketone bodies may bind to HCAR2 to inhibit glycolysis, which decreases energy substrate ATP below the cell membrane and finally activates the K⁺ ATP channel. In addition, ketone bodies simultaneously activate mitochondria and synthesize ATP by using oxygen to increase the total amount of ATP. That is, ketone bodies decrease ATP below the membrane and increase ATP in mitochondria, finally increasing the total amount of ATP in neurons. Since cells have highly distinctive compartments, they sometimes need to be mixed. By these two actions, ketone bodies can activate the K⁺ ATP channel to stabilize the membrane, leading to therapeutic effects against epilepsy.

11.4.2   A Big Advantage of the Ketogenic Diet

There is no other therapeutic option but to increase ketone bodies by changing to a ketogenic diet to combat refractory epilepsy. The author argued that most of the health effects of ketone bodies are possible when increased to a concentration of 0.5 mM. However, the K⁺ ATP channel is an exception. A ketone body concentration of at least 2 mM, ideally 5 mM, is required for K⁺ ATP channel activation. A ketone body concentration of 2 mM considerably inhibits the K⁺ ATP channel; 5 mM causes complete inhibition [4, 5].

It is a ketogenic diet that tends to increase ketone bodies to 5 mM. The ketogenic diet needs complete restriction of carbohydrate intake, but it is very tough to maintain this for a long time. However, it will have great effects if used properly in the right situation. It is a highly effective method to convert the major energy substrate from glucose to ketone bodies and completely inhibit refractory epilepsy. However, in my experience, this eating habit is a tough way to increase ketone bodies to 5 mM (from normal 0.1 mM). For example, the intake of much fat, as well as the complete removal of carbohydrates, is required. It is noted that strong efforts will be needed to keep ketone bodies at these levels [2].

11.5 Ketogenic Diet

11.5.1 The Basic Concept of the Ketogenic Diet

I will provide a scientific explanation for the ketogenic diet. Here, I do not mention of actual menu but instead the basic concept of a ketogenic diet. Everybody has learned three major nutrients in junior high school. Carbohydrates, fat, and protein are the major human nutrients. Based on common sense, the ideal diet has the following ratio:

Carbohydrate:fat:protein = 60:20:20

The standard Japanese diet has this composition. (I think the proportion of carbohydrates is too high.)

However, the ideal ketogenic diet ratio is:

Carbohydrate:fat:protein = 0:80:20

To try the ketogenic diet, the ketone ratio is the most important criterion for eating habits.

Ketone ratio = fat: (carbohydrate + protein)

The ketone ratio is regarded as the ratio of fat to other nutrients. In the ketogenic diet, the ketone ratio should be kept between 3:1 and 5:1. To keep the ketone body concentrations >5 mM, the ketone ratio should be at least 4:1.

In conclusion, to combat refractory epilepsy, the ketone ratio should be kept at 4:1.

In addition, there is another condition to the ketogenic diet.

Carbohydrate intake is strictly limited to less than 5 g/day.

This is the reason why most people cannot sustain this habit. Rice, bread, noodles, sweets, fruit, and most vegetables are not recommended in the ketogenic diet. The ketone ratio of a normal diet is 1:4 and that of the ketogenic diet is 4:1. You can imagine how tough to keep this diet for a long time.

11.5.2 History of the Ketogenic Diet

The concept of the ketogenic diet is around 100 years old. In the early twentieth century, it was known that fasting had a therapeutic effect on epilepsy. Interestingly, fasting has a highly therapeutic effect on most neurological disorders, including dementia and depression. Since fasting cannot be sustained for a long time, food that has the same effects was invented and called the "ketogenic diet." Around 1920, Russell Wilder invented this diet and reported that it had a big therapeutic effect against refractory epilepsy [2, 7].

As the ketogenic diet requires many restrictions and is difficult, only a small proportion of clinicians use it to treat epilepsy. In addition, a series of anti-epilepsy drugs have been developed over the past 80 years. If it is not refractory epilepsy, seizures can be controlled by these drugs. Consequently, the ketogenic diet has been neglected until recently. Further, in the past 30 years, more potent anti-epilepsy drugs that suppress seizures have become available. For example, GABA activators (tranquilizers) are developed to activate inhibitory neurons and may have partial inhibition of excessive action potentials. However, evidence has accumulated to show that even these potent drugs have no therapeutic effect. In contrast, they only cause serious side effects in refractory epilepsy [6].

Recently, as many researchers reported the therapeutic effects of ketone bodies in the brain, the use of a ketogenic diet has been reevaluated in neurological research. (Owing to developments in science and technology, the powers of ketone bodies can be examined. Maybe it is because of some scientists and doctors who have been steadily pursuing this research.) In addition, the ketogenic diet will likely attract much more attention in the future, not only for epilepsy but also to combat cancer and dementia [8, 9].

The ketogenic diet intends to have fat as the main energy substrate instead of carbohydrates to increase ketone body concentration. The increase in ketone body concentration prevents epileptic seizures and protects the cognitive ability of patients with dementia. In addition, it may prevent cancer progression. Studies on ketone bodies and the ketogenic diet are increasing steadily.

11.6 Anticancer Effects of the Ketogenic Diet

11.6.1 Adjuvant Effects of Ketone Bodies Against Malignant Tumors

Recently, many clinical tests on the ketogenic diet have been performed in major university hospitals worldwide. The application of the ketogenic diet in the neurological field is for the inhibition of brain tumors. It is the brain tumor that astrocytes are transformed into tumor cells. Most gliomas are malignant and, if the position is critical, there are no effective treatments. In addition, the brain tumors are inside the BBB, so most anticancer drugs cannot be used. The current situation is that irradiation and operation may be possible if the tumor is in a favorable position [8, 9].

Surprisingly, the ketogenic diet has a therapeutic effect on brain tumors (Fig. 7.6). The ketogenic diet itself has very small effects, but in combination with irradiation, it induces a high therapeutic effect. This is known as the "adjuvant effect." However, even in this case, there is a big problem that restrictions on carbohydrates must be maintained. Nevertheless, patients and their families may take some hope. For over a hundred years, researchers have known that ketone bodies can suppress the proliferation of cancer cells. Cancer cells have altered energy metabolism, completely different from normal cells. In the early twentieth century, Otto Warburg, a German biochemist, discovered these effects and they are known as the "Warburg effect" [10, 11].

Strangely, malignant cells have completely suppressed mitochondria. Thus, these cells cannot effectively produce energy and are completely dependent on ineffective glycolysis for energy production. Cancer cells gulp down glucose. As ketone bodies can be metabolized only in the mitochondria, it can be imagined that cancer proliferation would be suppressed. This is because ketone bodies cannot be used as an energy substrate by cancer cells. Indeed, ketone bodies can prevent cell growth. Although many papers are reporting their effects, ketone bodies have a very small effect, but they may delay cancer growth [10, 11].

Radiation alone had only a slight life-prolonging effect on human glioma transplanted into the mouse brain. When mice are fed a ketogenic diet, irradiation leads to potent inhibition of tumor growth. Surprisingly, most of the mice did not die within the experiment (up to 250 days) (Fig. 11.3b). In addition, there was a major effect on glioma, which is known as a highly malignant tumor. Please note that Fig. 11.3a shows 100 days, but Fig. 11.3b shows 250 days. Thus, the adjuvant effects of the ketogenic diet were enormous. These results suggest that ketosis induced by the ketogenic diet will have excellent adjuvant effects [10].

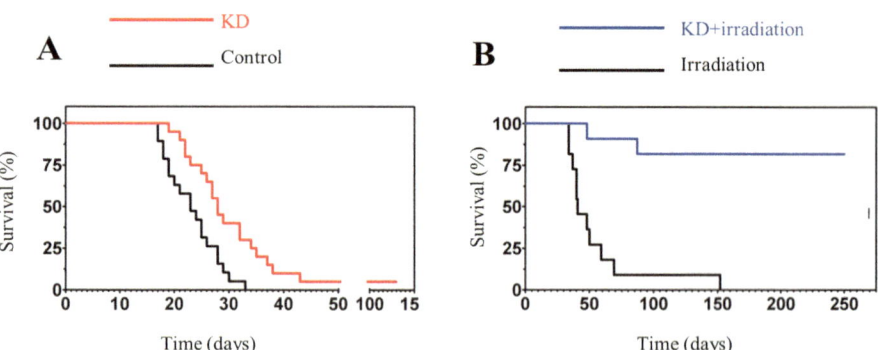

Fig. 11.3 Adjuvant effects of ketonic diet (KD) in mice [10]. KD+ irradiation induces long survival, suggesting that ketone bodies may extend survival, suggesting that KD enhances the anticancer effects of irradiation. This effect is termed an "adjuvant effect"

Combinations of several anticancer drugs should have bigger therapeutic effects, but patients will experience greater side effects. Food ingredients without side effects are required now to confer adjuvant effects to standard cancer therapy. Consequently, the ketogenic diet may be a possible food composition that has a major adjuvant effect. This cannot be replaced by other drugs and food ingredients. A recent report showed that the anti-cancer effects of the ketogenic diet may be in the case of human cancer patients [12]. The long-term outcomes of patients with Stage IV cancer who were treated with a ketogenic diet in Osaka University Hospital in Japan were reported [13]. A ketogenic diet continued for more than 12 months may significantly improve the survival rate of the patients. However, further research will be needed to conclude.

References

1. Perucca P, Scheffer IE, Kiley M. The management of epilepsy in children and adults. Med J Aust. 2018;208(5):226–33.
2. Sampaio LP. Ketogenic diet for epilepsy treatment. Arq Neuropsiquiatr. 2016;74(10):842–8.
3. Pan JW, Rothman TL, Behar KL, Stein DT, Hetherington HP. Human brain beta-hydroxybutyrate and lactate increase in fasting-induced ketosis. J Cereb Blood Flow Metab. 2000;20(10):1502.
4. Sada N, Lee S, Katsu T, Otsuki T, Inoue T. Epilepsy treatment. Targeting LDH enzymes with a stiripentol analog to treat epilepsy. Science. 2015;347(6228):1362–7.
5. Mollajew R, Toloe J, Mironov SL. Single KATP channel opening in response to stimulation of AMPA/kainate receptors is mediated by Na+ accumulation and submembrane ATP and ADP changes. J Physiol. 2013;591(10):2593–609.
6. Harris N, Baba M, Mellor C, Rogers R, Taylor K, Beringer A, Sharples P. Seizure management in children requiring palliative care: a review of current practice. BMJ Support Palliat Care. 2020;10(3):e22.
7. Wheless JW. History of the ketogenic diet. Epilepsia. 2008;49(Suppl 8):3–5.
8. Puchalska P, Crawford PA. Multi-dimensional roles of ketone bodies in fuel metabolism, signaling, and therapeutics. Cell Metab. 2017;25(2):262–84.
9. Grabacka M, Pierzchalska M, Dean M, Reiss K. Regulation of ketone body metabolism and the role of PPARα. Int J Mol Sci. 2016;17(12):2093.
10. Vaupel P, Multhoff G. Revisiting the Warburg effect: historical dogma versus current understanding. J Physiol. 2021;599(6):1745–57.
11. Schwartz L, Supuran CT, Alfarouk KO. The Warburg effect and the hallmarks of cancer. Anticancer Agents Med Chem. 2017;17(2):164–70.
12. Abdelwahab MG, Fenton KE, Preul MC, et al. The ketogenic diet is an effective adjuvant to radiation therapy for the treatment of malignant glioma. PLoS One. 2012;7(5):e36197.
13. Egashira R, Matsunaga M, Miyake A, Hotta S, Nagai N, Yamaguchi C, Takeuchi M, Moriguchi M, Tonari S, Nakano M, Saito H, Hagihara K. Long-term effects of a ketogenic diet for cancer. Nutrients. 2023;15(10):2334.

Epilogue

The Blue Bird

Please imagine that an explorer travels all over the world to find a bluebird that stops aging with its cry. He crossed the sea and climbed over mountains and was about to die in the desert. Finally, he gave up on finding it and returned home to Tokyo. He found a beautiful bird in the small cage through a lace curtain near the window. He put his knee on the floor while laughing and began to cry with tears running down his cheeks. He said, "He has been here," his heart and body were healed by a beautiful cry. Does my writing overstate that the ketone bodies are like bluebirds for us? Over the years, various bad rumors about ketone bodies have pushed it out of favor of public opinion. During the serious crisis of ketoacidosis, the body may make ketone bodies to rescue it from the crisis. However, ketone bodies have been regarded as dangerous and disliked by clinicians. This is a very sad story. The true criminal here is that which induces the crisis.

It may be the unexpected near future when ketone bodies are acquitted due to insufficient evidence. Where the brain is concerned, I wish to declare, with strong confidence, that ketone bodies are innocent. While many people condemn ketone bodies as the cause of ketoacidosis, an increasing number of people pursue health and antiaging strategies by following an extreme carbohydrate-free diet to increase ketone body concentrations. Extreme restriction of carbohydrates is not only spreading throughout Europe and the USA, but also to Japan for various reasons, such as diet, building muscle, and treating disease. Through these trends, it is broadcasted that many people eat "Neta," instead of rice in Sushi restaurants.

From a historical perspective, people with extreme views may take the initiative from public opinions during the transition of values. However, these extreme opinions often cause a series of conflicts against the current cultures and traditions. Owing to these conflicts, these extreme concepts lose favor within a short time. In some cases, these conflicts may invade the daily life of couples. Not only eating

T. Satoh, *Hybrid-Powered Brain*, https://doi.org/10.1007/978-3-031-54150-6

habits but also food color and design, are changed. This is why differences between couples are often a frequently discussed topic. If they can choose an intermediate, hybrid system and if such an eating habit may be ideal, everyone can take a break and gain peace of mind. This book is required to show that such a hybrid system is reasonable from a scientific perspective. This is why the author has proposed "the small ketogenic" to take advantage of ketone bodies without any conflicts with current eating habits. I do not agree with a complete vegetarian diet or with extreme carbohydrate restriction. This is the basic concept of this book.